A Cartoon Primer
of Modern Mathematics

A Cartoon Primer
of Modern Mathematics

Chih C. Yang

Willow Creek Science Institute

2017

Copyright © 2017 by Chih C. Yang

All rights reserves. This book or any portion thereof may not be reproduced or used in any manner whatsoever without the express written permission of the publisher except for the use of brief quotations in a book review or scholarly journal.

First Printing: 2017

ISBN 978-0-9992362-0-8

Willow Creek Science Institute
P.O. Box 1847
Buies Creek, NC 27506

CONTENTS

Preface ix

1. **What is Modern Mathematics?** 1
   - A Brief History of Mathematics 2
   - The Middle Ages 4
   - The Renaissance in Italy 6
   - The Invention of Calculus 8
   - Mathematics in the 18-19 Century 9
   - Mathematics in the 20th Century 10

2. **The Science of Patterns** 13
   - The German Tank Problem 14
   - Bomber's Bullet Holes Pattern 16
   - Criminal Patterns 17
   - Mining the Big Data 19

3. **Abstraction and Thought Experiments** 23
   - Abstraction
     - Why Abstraction? 24
     - Concrete vs. Abstract Thinking 25
     - What is Abstraction? 26
     - Levels of Abstraction 32
     - Application: The Model-T Ford Assembly Line 35
   - Thought Experiments
     - Plato's Allegory of the Cave 37
     - Little Johnny's Thought Experiment 39
     - Galileo's Sailing Ship 41
     - Einstein's Train 43
     - Einstein's Twin 45
     - Applications 46

4. **Set and Infinity** 49
   - Why Set Theory? 50
   - Infinite Sets 52
   - How Big is Infinity? 54
   - Counting an Infinity Set 55
   - Hilbert's Hotel 58
   - Real Numbers 61
   - Paradox in Set Theory 65

5 Mathematical Induction 69
    Carl Friedrich Gauss 70
    The Principle of Mathematical Induction 73
    Tower of Hanoi 76
    The Tromino Puzzle 79

6 The Structure of Algebra 85
    The Beginning of European Algebra 86
    The Rise of European Algebra 87
    The Anatomy of Algebra 89
    Function 90
    Binary Operation 95
    The Order of Operations 100
    Relation 101
    Properties of Relations 106

7 Modular Arithmetic 113
    The Greatest Common Divisor 114
    Euclidean Algorithm 115
    Euclidean Algorithm and the Golden Ratio 118
    Congruence of Integers 120
    Modular Arithmetic in Everyday Life 122
    Applications 125

8 Chinese Counting 133
    Counting 134
    Chinese Remainder Counting 135
    Mathematical Description 139
    Pirates of the Carolinas 141
    Applications
        Computer Precision Arithmetic 142
        Cryptography 144
        RSA Public-key Cryptosystem 146

9 The Quest for Symmetry 153
    The Evolution of Scientific Methods 154
    Galileo's Experiments and Mathematics 155
    Isaac Newton's Giant Step 156
    The Breakdown of Classical Physics 157
    Aesthetics 159
    Symmetry 161
        Groups of Symmetry 162
        Three-dimensional Symmetry 166
        Chirality 167
        A Tragic Example of the Importance of Chirality 169
        Symmetry in Physics 170
        Symmetry in the Sky 171

    Symmetry in Wealth Management 172
    Symmetry in Music 173
    Symmetry in Image Recognition 174

10 Algebraic Systems 177
    Algebraic Systems 178
    Group, Ring, Integral Domain and Field 180
    Group 181
    Classification of Groups 188
    Non-commutative Operations 189
    Applications of Non-commutativity 195
    Isomorphism 197
    Homomorphism 200

Appendix
    A Pattern from the Finite Differences 205
    A Well-defined Binary Operation 206
    Bomber's Bullet Holes Pattern 207
    Pirates of the Carolinas 208
    RSA Public-key Cryptosystem 209

Index 211

A Ferry to the Land of Abstract Math

Preface

A Picture of Educational Stagnation

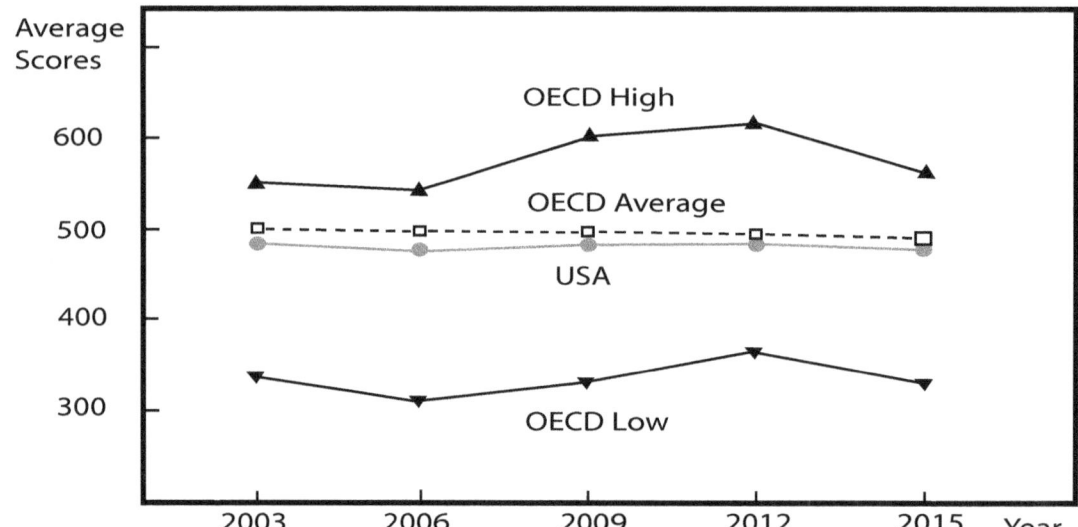

Average scores of 15-year-old students on the PISA mathematics literacy tests from 2003 to 2015

The Programme for International Student Assessment (PISA) is an Organization for Economic Co-operation Development (OECD) initiative that aims to evaluate the scholastic performance of students worldwide.

According to tests administrated to 15-year-old students by PISA, American students are below the average of the 72 countries studied in mathematics, far behind teens in Asia. In 2013 the then US Education Secretary, Arne Duncan described the PISA results as "a picture of educational stagnation."

Mathematics is Composed of Thinking and Computing Skills.

The general public's images of mathematics has been restricted to numbers and computing methods. However, mathematics is more than just that. At its core, mathematics is about thinking and conceptualizing using a process of logic and abstract reasoning.

Today, American's mathematics education is shifting towards reasoning and conceptualizing. Yet, computing skills still remain important as reasoning is built on a foundation of solid computing skills. They are closely intertwined.

Mathematics Education in Asia

Asian students in a test prep center

A country's educational system reflects the values of its society.

■ For most Asian students, a successful college entrance examination is the ultimate goal.

■ In countries like China, Japan, Singapore, South Korea, and Taiwan, this exam determines the course of a student's career and life. Primarily, these tests are standardized and focus on computing skills. Local test-prep centers offer demanding courses to assist in preparing and are popular among students during after-school hours. Test results are prioritized over practical knowledge and ability.

■ Because this type of educational system emphasizes cook-book style computational skills, labour-intensive manufacturing sectors are largely benefitted in these Asian countries.

Mathematics Education in the U. S.

From the late 1940s to the early 1990s, American society was defined by the Cold War. With the Soviet Union's successful launch of Sputnik in 1957, Americans awoke to a new frontier as the great Space Race and Arms Race began.

At the onset, US policy makers blamed the lag on the country's subpar science and mathematics programs. They believed the curricula and teaching methods were outdated and needed to be overhauled. As a result, the "New Math" movement was born.

The New Math Movement of the 1960s

In the 1960's, math scholars and educators began emphasizing thinking skills in mathematics curricula. No longer would math textbooks solely emphasize computational skills and applications. Now, students would learn the structure of mathematics, reasoning and conceptualizing.

Johnny can't do math.

New Math:
"It's the method that's important, never
*   mind if you don't get the right answer."*
- Tom Lehrer

While the reform was implemented for several years, the effect was largely unknown. Eventually, the movement died out as educators took the movement to an extreme by shifting the focus to only conceptualization while ignoring computational skills.

By the mid-1970s, the movement had completely faded out.

America's Math Wars - Math Education Reform

America's discontent with math education kept reform movements and controversies alive. In the 1990s, the debate over math education and curricula rejuvenated with the publication of the Curriculum and Evaluation Standards for School of Mathematics by the National Council of Teachers of Mathematics. The main argument centered over whether a traditional approach or "reform math" approach was better. The reform math approach emphasizes students' discovering their own knowledge and conceptual thinking. The "wars" died down by the mid-2000s.

The Math Wars Rage On

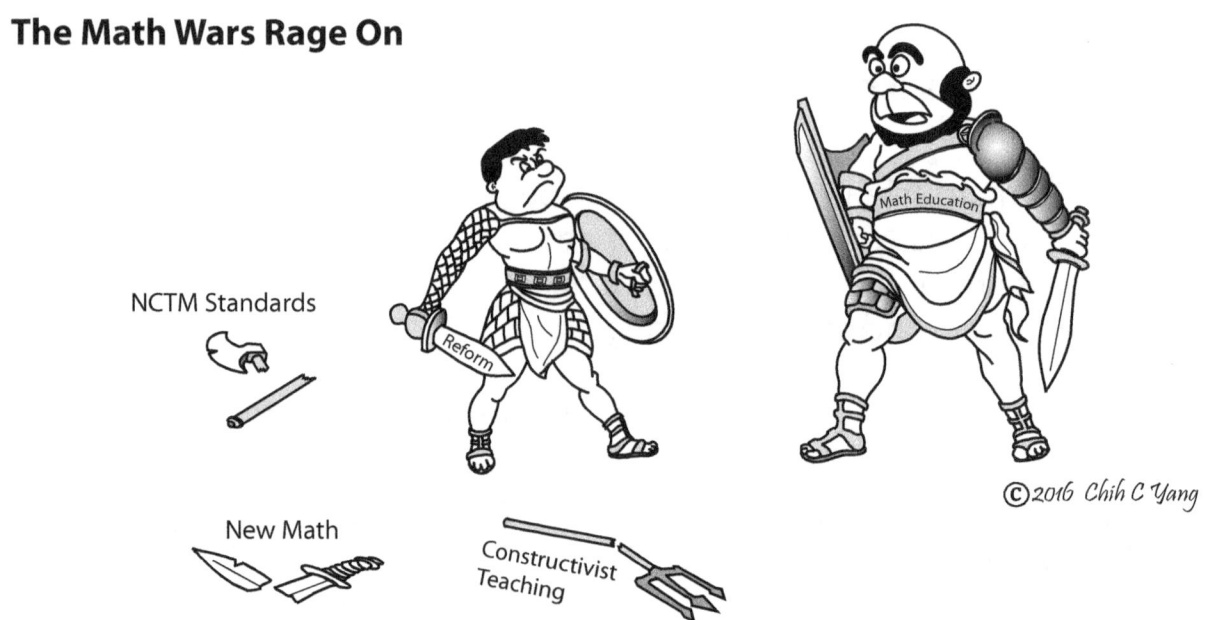

With the implementation of the Common Core State Standards in 2010, the math reform battles flared up again.

Where are We?

Today, the "thinking" approach to math (i.e. emphasizing thinking and conceptualization) dominates mathematics in college at the post-calculus level, while the K-12 math education still struggles for directions.

Abstract Math at the University Level

At the post-calculus university level, the primary emphasis is on understanding structures and abstract concepts.

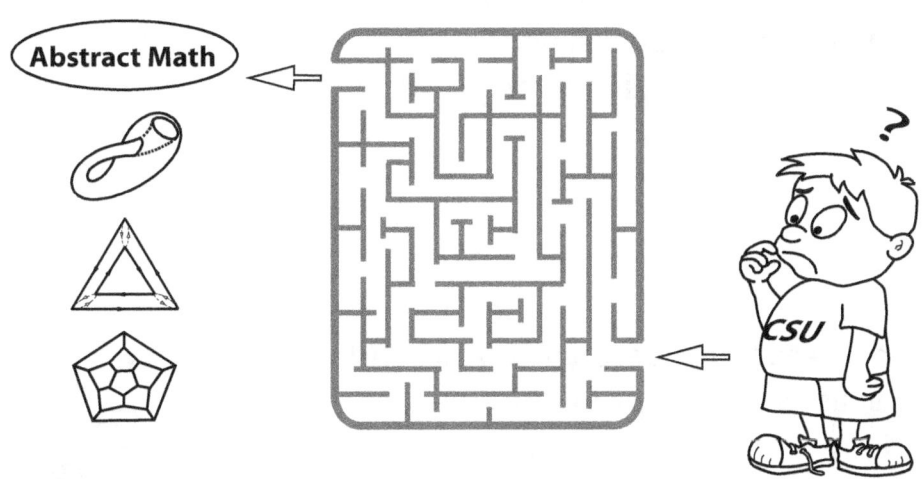

Reaching this level of thinking has proven difficult as relatively few students are well-equipped to think in this manner. In order to prepare students for abstract mathematics, many transitional books are available. However, these transitional books primarily focus on the treatment of logic and the art of proof writing. In many cases, this approach intimidates students causing them to shy away from abstract mathematics.

Objectives of This Book

This book is intended for readers of varying mathematical abilities with the goal of helping readers to appreciate and enjoy the beauty of modern abstract mathematics and to pique their interest in learning additional mathematical topics and ways of thinking. In order to keep the reader engaged, the author has attempted to use as many concrete examples as possible to illustrate the features of general concepts.

Acknowledgements

I wish to express my deepest gratitude to the following people for their helpful comments, suggestions, and edits in this endeavor (in alphabetical order).

Mark Evans, Cultural Conservation Foundation of North Carolina
Ronald Fulp, N. C. State University
William Johnson, N.C. State University
Gene Lowrimore, Decision Systems; former Senior Scientist, Duke University
Ronald Sneed, N.C. State University; Major General (retired), US Army
Louise Taylor, Meredith College, Raleigh, NC
James and Beverly West, Cornell University

Finally, I would like to thank Carolyn, Jo Ann, and Matthew for their assistance in preparing the manuscript and my wife, Yu-Mong, for her support and encouragement.

1
What is Modern Mathematics?

A Brief History of Mathematics

Up until the 6th-5th centuries B.C., mathematics was the study of numbers and basic arithmetic.

Sexagesimal (base-60) system

Babylonian

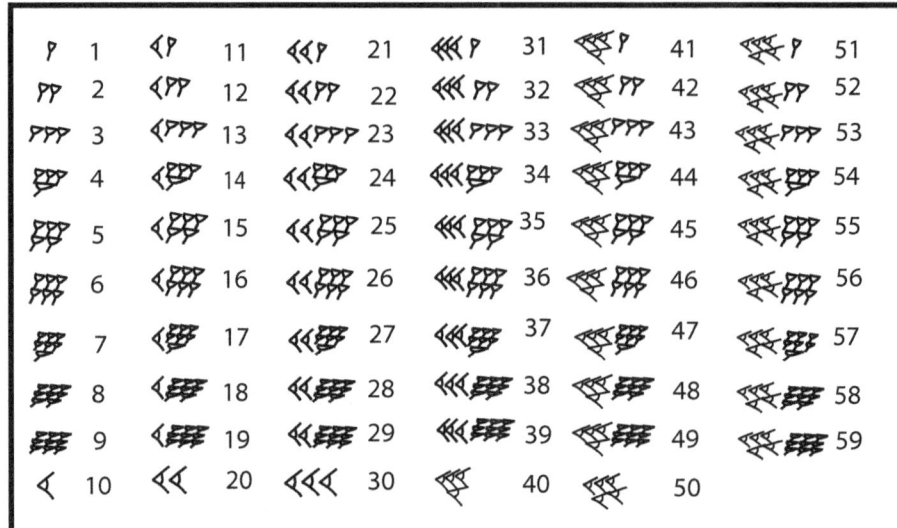

Up to date, we still use the sexagesimal system. For example, there are 60 minutes in an hour.

Egypt

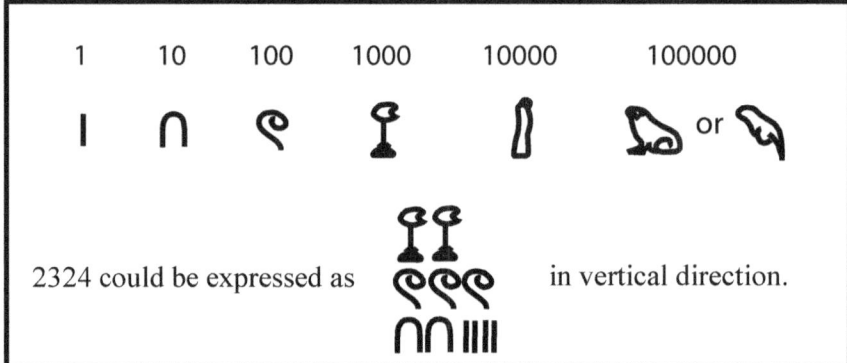

2324 could be expressed as ... in vertical direction.

Maya

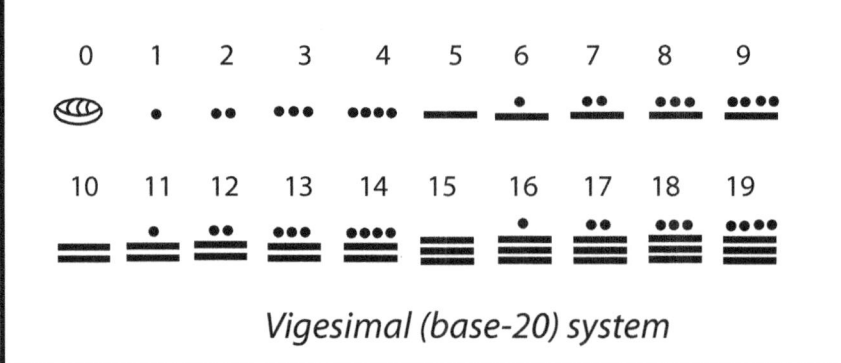

Vigesimal (base-20) system

The concept of zero appeared very early, but did not spread to other civilizations outside the Mayans.

From 500 BCE to 300 CE, the Greeks made significant advances in mathematics, especially geometry.

For the Greeks, mathematics was not just utilitarian. Mathematics was also an intellectual pursuit having both aesthetic and religious elements. Euclid's geometry is the first complete deductive science in history.

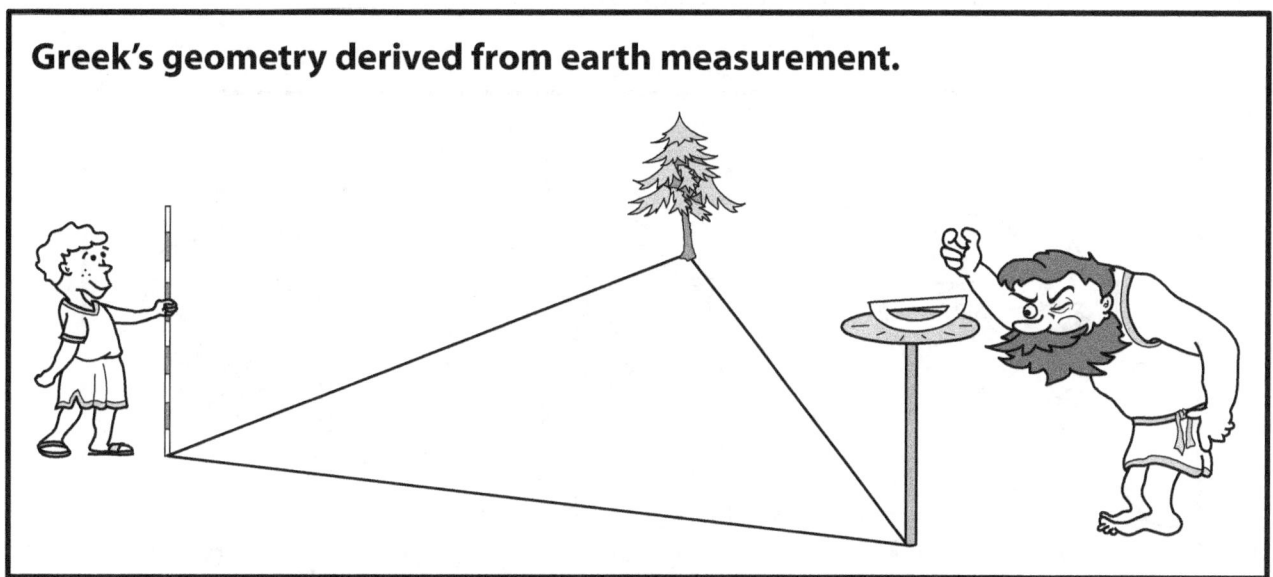

Greek's geometry derived from earth measurement.

Unsolved Geometry Problems

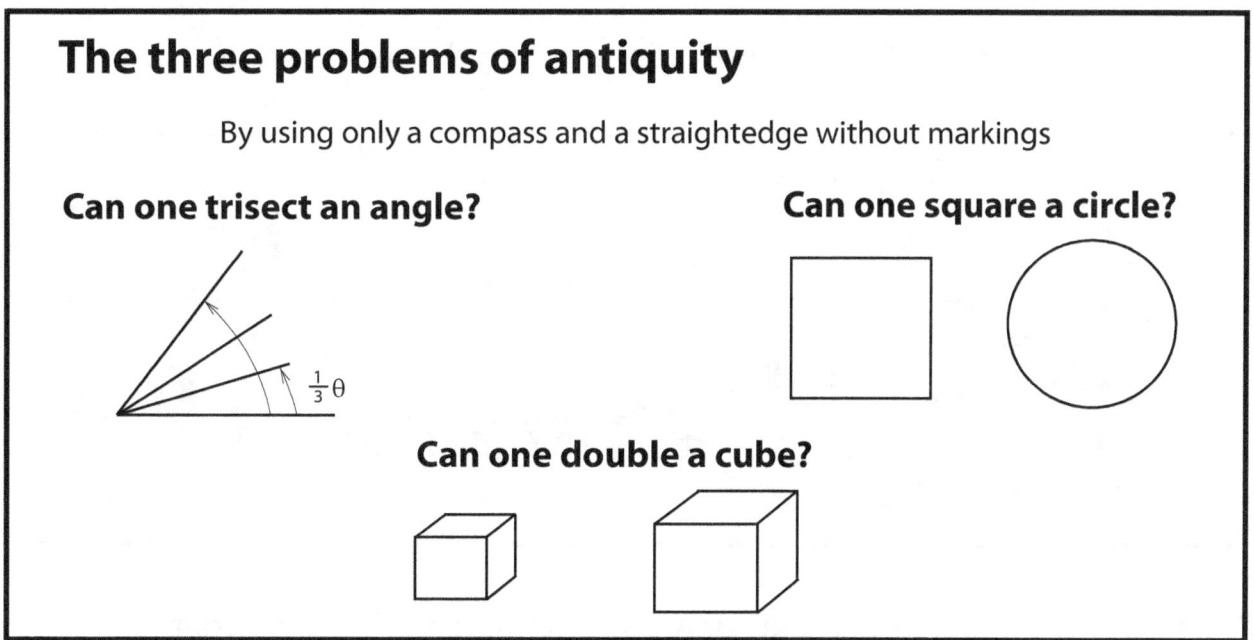

The three problems of antiquity

By using only a compass and a straightedge without markings

Can one trisect an angle?

Can one square a circle?

Can one double a cube?

They had vexed mathematicians for over 2000 years until it was shown by algebra that they could not be constructed with compass and straightedge.

The Middle Ages (5th - 15th centuries CE)

After the fall of the Rome Empire, European civilizations were weakened by plagues, wars, and barbarian invasions. For nearly a thousand years, the development of mathematics in Europe came almost to a standstill.

Conversely, some Eastern civilizations during the period were thriving.

Hindu-Arabic Numerals

■ In the 12th century, many texts of Arabic mathematics were translated into Latin. Among them was the **Algebra** of al-Khwarizmi.

- The Hindu-Arabic numerals were also brought to Europe, including the number zero; however, "zero" was regarded with skepticism until the 16th century.

The number zero originates in India during the ninth century.

- Negative numbers were known but not fully accepted in Europe. They were referred to as "absurd numbers."

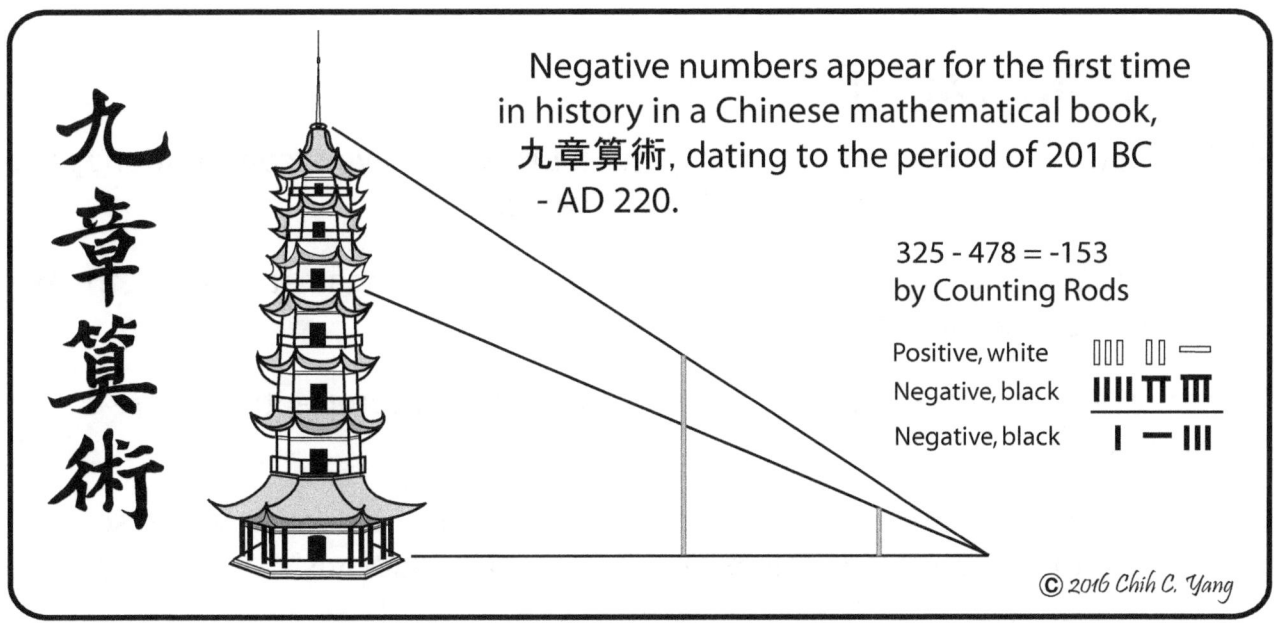

Negative numbers appear for the first time in history in a Chinese mathematical book, 九章算術, dating to the period of 201 BC - AD 220.

325 - 478 = -153
by Counting Rods

Positive, white
Negative, black
Negative, black

The Renaissance in Italy
- The rise of European Algebra in the 16th century

The first show of $\sqrt{-1}$

- During the mid-16th century in Italy, math problem solving competitions were a popular pastime and a way to display talents.

- Mathematical discoveries were kept secret to be used against other mathematicians in some contest.

- Heavy bets were made. For a good contestant, these competitions were a source of income.

- Girolamo Cardano (1501 - 1576) experimented with the number $\sqrt{-1}$ to solve math problems, but ultimately believed it was "nonrigorous".

Algebra emerged as a major branch of mathematics

The great accomplishment of 16th Century algebra

- Finding a general solution for cubic and quartic equations

$$x^3 + ax^2 + bx + c = 0$$
$$x^4 + ax^3 + bx^2 + cx + d = 0$$

- ■ The challenge was to discover the next step to finding the general solution to equations of degree 5 or higher.

- ■ During the next 200 years, no one succeeded.

- ■ In the 19th century, Niels Henrik Abel (1802 - 1829) studied algebra's structure and concluded: There is no formula for the roots of an equation with a degree of 5 or higher.

- ■ This conclusion signaled the beginning of modern algebra.

The Invention of Calculus

Calculus was developed independently in the 17th Century by Isaac Newton (1642 -1727) and Gottfried Wilhelm Leibniz (1646 - 1716).

A new period of mathematics began. The emphasis of mathematical interests shifted to the domain of variable (or changing) quantities.

The Difference between Calculus and Previous Elementary Mathematics

Previous mathematics had been largely restricted to counting, arithmetic calculation, geometry and algebra.

Calculus studies continuous motion and changes.

Continuous motion and changes

A Rolling Circle - Cycloid

A problem was posed by Johann Bernoulli in 1696 to challenge other mathematicians. The problem was to determine the shape of a wire down which a bead might slide from point **A** to point **B** in the shortest **time**, when B is not directly below A.

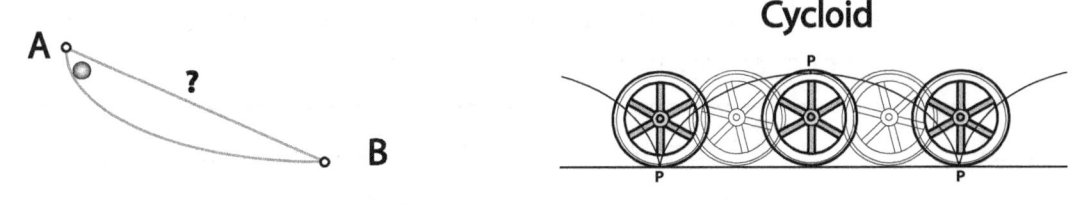

The answer is half of an arch of one inverted cycloid, rather than a straight line.

Mathematics in the 18th - 19th Century

From the middle of the 18th century there was an increasing interest in mathematics itself, not just its applications. By the end of the 19th century, mathematics had become the study of number, shape, motion, change, space, its own structures, and the abstraction and logical reasoning of mathematics.

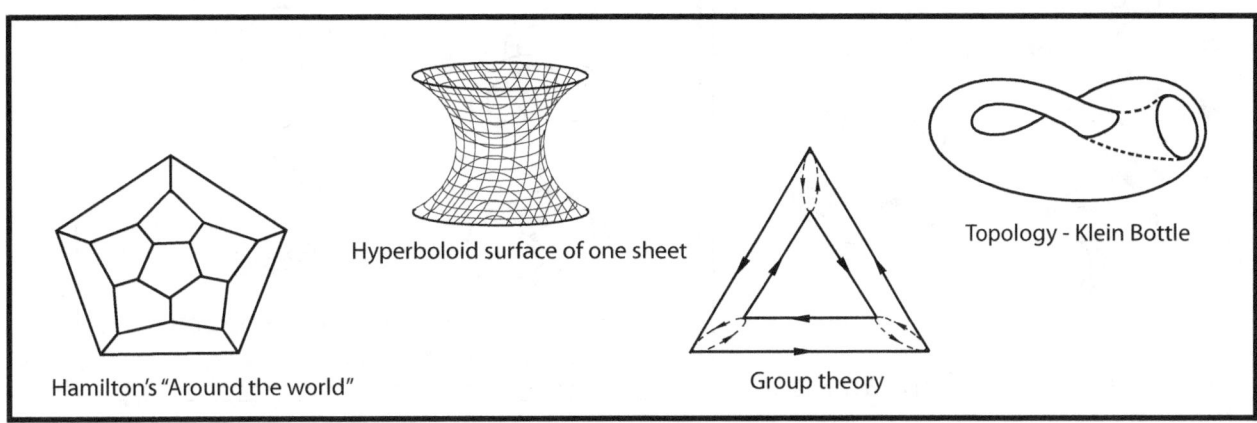

Hamilton's "Around the world"

Hyperboloid surface of one sheet

Group theory

Topology - Klein Bottle

Mathematics in the 20th Century

The development of mathematics in the last century was dramatic. Along with the further development of earlier fields, many new branches of mathematics have been explored.

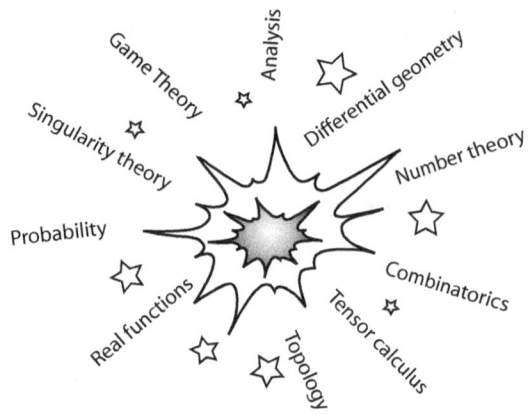

Given that explosion of activity, a question arises:
What is mathematics?

Until the 1980s, most mathematicians agreed that **mathematics was the science of patterns.** Mathematicians seek out abstract patterns in numbers, shapes, space, science, computers, nature, life, behavior, etc.

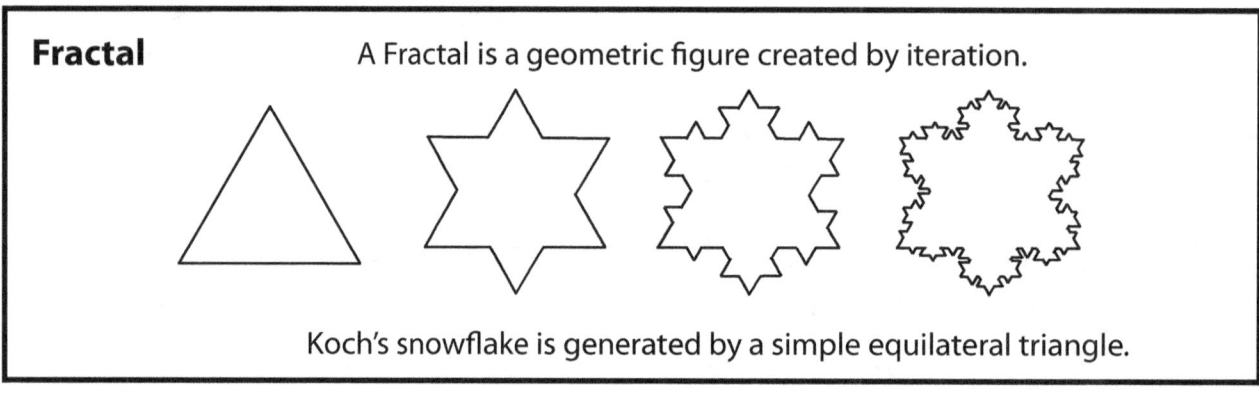

Fractal — A Fractal is a geometric figure created by iteration.

Koch's snowflake is generated by a simple equilateral triangle.

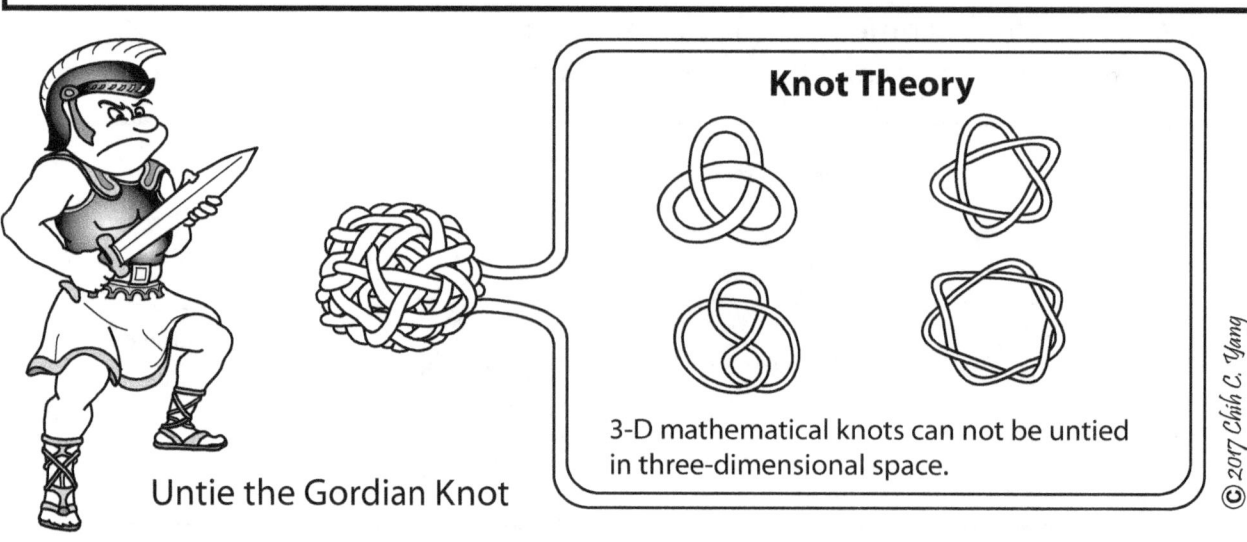

Untie the Gordian Knot

Knot Theory

3-D mathematical knots can not be untied in three-dimensional space.

References – Modern Mathematics

[1] Anton, Howard, Calculus with Analytic Geometry, 4th Edition. John Wiley & Sons, 1992

[2] Berlinghoff, William P.; Mathematics – The Art of Reason, D.C. Heath and Company, Boston, 1968

[3] Boyer, Carl B.; Merzbach, Uta C., A History of Mathematics, 2nd Edition. John Wiley & Sons. 1991.

[4] Chinn, W. G., and Steenrod, N. E., First Concepts of Topology: the geometry of mappings of segments, curves, circles, and disks. Random House/The L. W. Singer Company, NY. 1966.

[5] Devlin, Keith J., Sets, functions, and Logic: an introduction to abstract mathematics, 3rd Edition. Chapman & Hall/CRC, FL. 2004

[6] Devlin, Keith J., Life by the numbers, John Wiley & Sons, Inc. NY. 1998

[7] Geometry – Integration, Application, Connections, Glencoe/McGraw-Hill, OH. 1998

[8] Mathematics. Encyclopedia of Mathematics. URL: http://www.encyclopediaofmath.org/index.php?title=Mathematics&oldid=23895. March 2012

[9] Nicodemi, Olympia E., Sutherland, Melissa A., and Towsley, Gary W., An Introduction to Abstract Algebra with Notes to the Future Teacher.Pearson/Prentice Hall, NJ. 2007

[10] Odifreddi, Piergiorgio; The Mathematical Century –The 30 Greatest Problems of the Last 100 Years, The Princeton University Press, 2004

[11] Stewart, Ian, Concepts of Modern Mathematics, Dover Publications. 1995

2
The Science of Patterns

Mathematics is the science of patterns. Mathematicians seek out patterns in numbers, shapes, space, time, science, computers, nature, behavior, etc. Geometry focuses on the pattern of shapes, whereas calculus studies the pattern of motion and change. Statistics focuses on the pattern of population and chance. Mathematics can also be the study of patterns of our daily lives. Here are some examples:

The German Tank Problem

The US M4 Sherman tank vs. the German Panther

The M4 Sherman tank was the primary tank utilized by both United States and some of its Allies during World War Two.

M4 Sherman tank, US 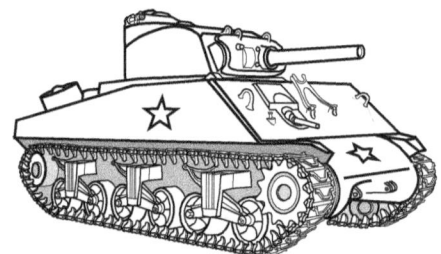   Panther tank, German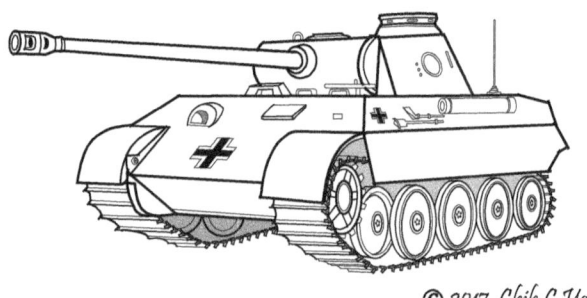

(Remix: Joost J. Bakker - M4 Sherman tank, CC BY 2.0, commons.wikimedia.org)

© 2017 Chih C Yang

The German Panther tank was superior to the American M4 Sherman in almost every respect.

In 1941 - 1942 during World War II, the Allies force had very little idea of the enemy's ability to produce the Panther tank. Prior to D-Day, the Allies were anxious to know the number of tanks being produced.

Information gathered from conventional intelligence and battlefield counts were contradictory and unreliable. As a result, mathematicians were called for assistance.

Statistical Estimator

- From the serial numbers of captured or destroyed German tanks, gearboxes and other components, statisticians discovered a **pattern** and figured out an estimator. From the formula, they estimated that the Germans produced 246 tanks a month.

- Post-war captured German production records showed the actual production rate to be 245 per month.

- In business applications, this method is used in auditing, credit check and market research. One example is the measure of the "cloud computing" market. Many companies maintain confidentiality with regard to company revenue and customer privacy. With this method, researchers decrypted the serial numbers from market leader, Amazon's virtual machines and came up with an estimate of the market size.

Bomber's Bullet Holes Pattern

■ During World War II the British bombed the Germans daily.

■ Wanting to improve the odds of bringing troops home, a mathematician named Abraham Wald studied the location of bullet holes in British planes and determined where to added armor plating.

Where would you add the armor plating to?

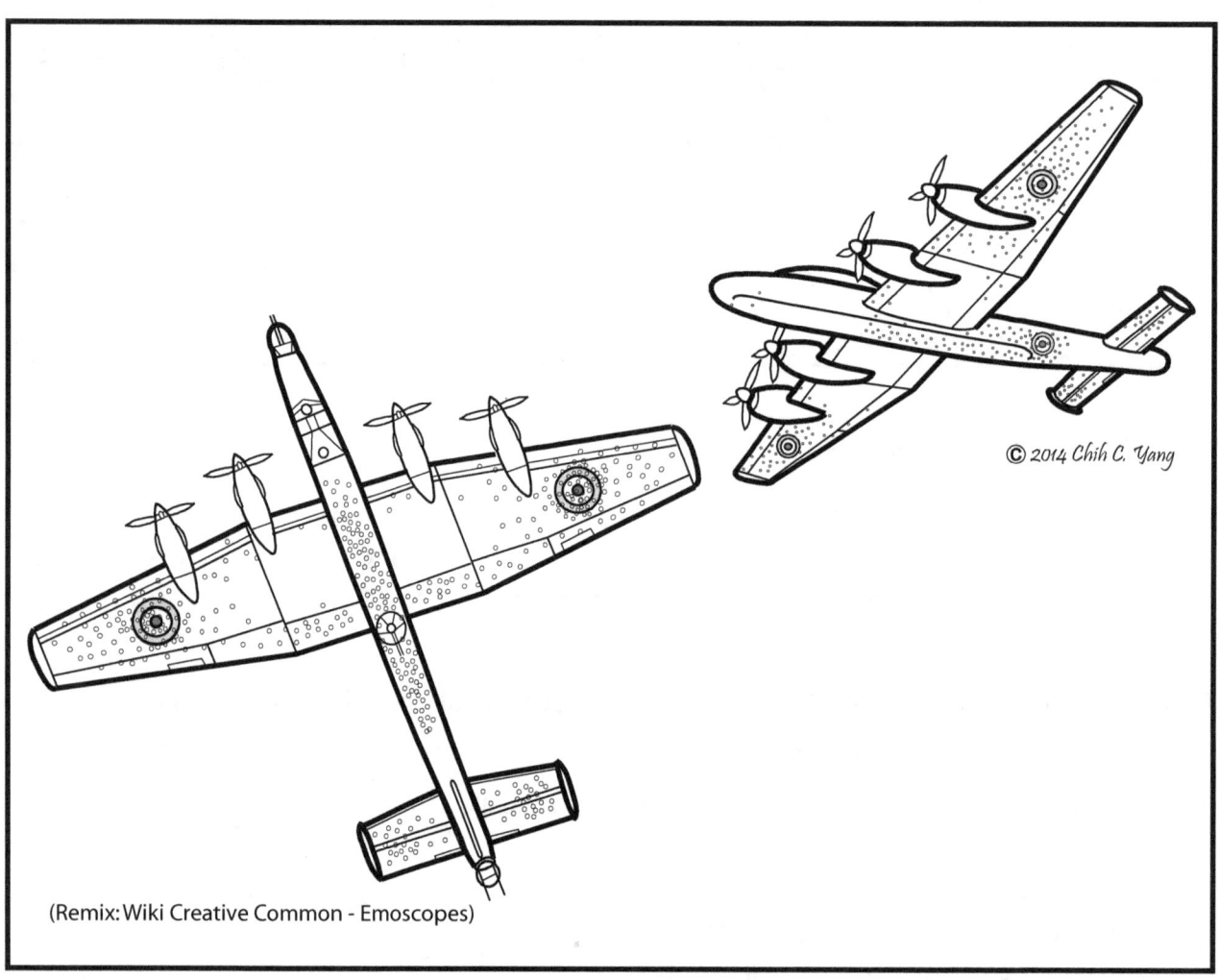

(Remix: Wiki Creative Common - Emoscopes)

Answer: Add the armor to the spots with no bullet holes.

(see appendix for details)

Criminal Patterns

At the US-Mexico border inspection station, SENTRI lanes allow qualified drivers to be searched less and pass through quickly.

The SENTRI drivers are the target of Mexico drug smugglers.

Drug smugglers paid teenagers to act as "lookouts". Their job was to monitor SENTRI drivers and find out their travel patterns.

Those with SENTRI access tended to be professionals and students with a consistent travel routine.

Once a vehicle was identified, the smugglers would follow the vehicle and plant a GPS tracker. They would copy the dashboard VIN number and have a locksmith create two keys for the car.

At night, marijuana would be sneaked into the trunk of the targeted vehicle and transported across the border the next day. On the US side, the smugglers would retrieve the drugs by using the second key.

Unsuspecting individuals would unknowingly smuggle drugs across the border. Several innocent people, including a school teacher, doctor, and college student, were arrested, detained or incarcerated.

The court and law enforcement noticed a pattern in these cases.

Detained suspects passed through the same SENTRI station.

All the suspects had no previous criminal records and maintained their innocence.

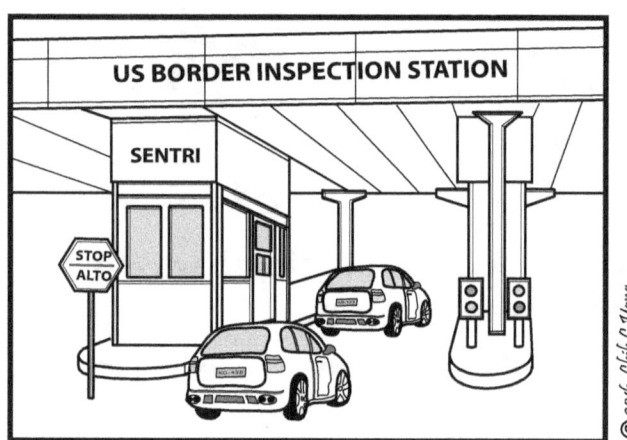

There were two duffel bags in each case.

Most suspects drove cars of the same make.

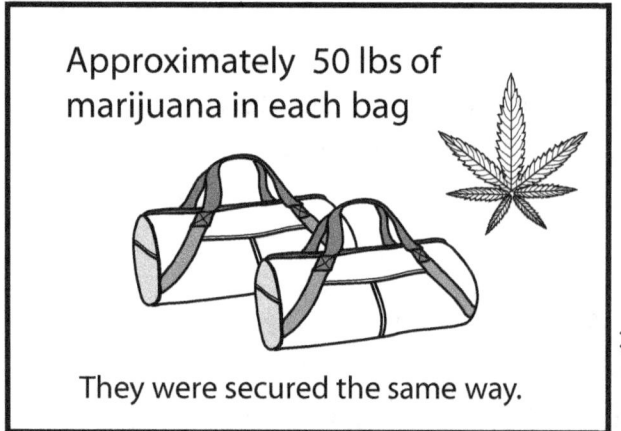

An investigation, led by the FBI, convicted two El Paso men in 2012 and brought justice for the victims.

Mining the Big Data - Finding Hidden Patterns

Data mining is an analytical process to search for hidden patterns and trends in large data sets. It provides new ways of looking at data.

As an interdisciplinary field of computer science and mathematics, data mining has been widely used in the development of new products, sales campaigns, customer retention and monitoring fraudulent activities within the banking, retailing, and telecommunication industries. Scientists also employ it as a research tool. The process of data mining involves artificial intelligence, database systems and mathematics, (i.e. sets, algebra, statistics, graphs, etc).

In the financial industry, one example where data mining is employed is the forcast of loan payments.

Factors related to the risk of loan payments include loan-to-value ratio, term of the loan, customer income, education, residence, credit history, etc. By analyzing the credit history, calculations determined the dominant risk factor is the payment-to-income ratio rather than the customer's personal profile such as education, residence, region, and income.

Mining Data in Bioinformatics

During the past decade, there has been an explosion of biomedical research. Most biomedical research focuses on DNA sequence analyses, which have led to the discovery of genetic causes for many diseases.

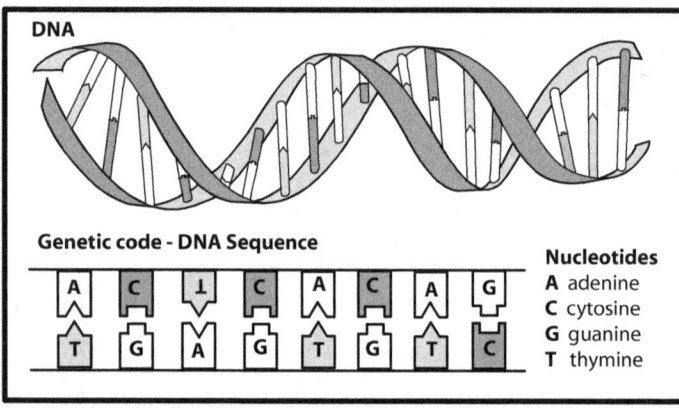

The patterns of DNA Sequences are descriptive rather than numerical.

However, decoding sequences is a huge undertaking. Human beings have about 100,000 genes. A gene is comprised of hundreds of nucleotides which are arranged in a particular sequence. There are unlimited ways in which nucleotides can be arranged to form a gene.

To identify a single candidate sequence for specific research is like trying to find a needle in the haystack.

Handling large volumes of data with non-numerical patterns is a big challenge. Fortunately, with advanced techniques like pattern analysis and data visualization, data mining is a powerful tool for researchers to decode DNA sequences.

A Pattern from Finite Differences

The Finite Difference Method, computes the rate of change, is a discrete analogy to the calculus derivative. Much of the methodology originated from Sir Issac Newton's work and remained popular with mathematicians for centuries. One important application of the method is found in numerical analyses and computational sciences.

A data-smoothing example is shown below:

> Example:
> One of the following numbers was misprinted, which one?
>
> **1   3   6   11   20   31   48   71   101**
>
> Hint: There is a pattern when you reach the fourth difference.

Can you find it?
(see appendix for answer)

References – The Science of Patterns

[1] Criminal Complaint; Case No. 11-3330-G; United States of America v. Jesus Chavez, Western District of Texas, United States District Court, July 1, 2011.

[2] David S. Moore; The Basic Practice of Statistics, 2nd Edition. W. H. Freeman and Company, New York, NY10010, 2000

[3] Evans, James R.; Business Analytics- Methods, Models, and Decisions, Pearson Education, 2013

[4] Han, Jiawei and Kamber, Micheline; Data Mining: Concepts and Techniques, 2nd Edition; The Morgan Kaufmann Series in Data Management Systems. Amsterdam: Elsevier, 2006

[5] Kantardzic, Mehmed; Data Mining: Concepts, Models, Methods, and Algorithms, 2nd Edition, Institute of Electrical and Electronics Engineers, John Wiley & Sons, 2011

[6] Ruggles, Richard and Brodie, Henry; An Empirical Approach to Economic Intelligence in World War II, The Journal of the American Statistical Association, Vol. 42, No. 237, pp. 72-91, March 1947

[7] Sawyer, W. W.; Introducing Mathematics: 3, The Search for Pattern, Penguin Books, Baltimore, Maryland, 1970.

[8] Rao, G Schanker; Numerical Analysis, New Age International, Daryaganj, Delhi, Ind. 2006

[9] Scheid, Francis; Theory and Problems of Numerical Analysis, Schaum's Outline Series in Mathematics, McGraw-Hill Book Company, New York, 1968

[10] Tank in the cloud, Information technology goes global, pp. 49, The Economist, January 1, 2011

[11] The FBI Press Releases; El Paso Man Sentenced to 20 Years in Federal Prison in Marijuana Smuggling Scheme, El Paso Division, Western District of Texas, U.S. Attorney's Office, September 11, 2012.

3
Abstraction and Thought Experiments

Why Abstraction?

A 2000-Year Old Question

There is an old story of one of Euclid's students who, upon learning the first theorem, proclaimed, "What do I get out of learning these things?" Euclid then called his servant and said, "Give him three pennies, since he must make gain out of what he learns."

The student's question is a common one; "Why do I have to study all this abstract stuff?" One answer is that someone found these abstractions useful, and that there are plenty of applications. However, there is something more important than applications.

Mathematics is a discipline that requires a lot of thinking. The need for abstract thinking permeates every branch of modern mathematics. However, abstract thinking is also essential to many other disciplines such as science, engineering, business, law, medicine and education. No discipline is better equipped to cultivate these skills than abstract math.

Concrete vs. Abstract Thinking

Examples

Concrete Thinking
- Facts & tangible objects

Abstract Thinking
- Deep thinking & general concept

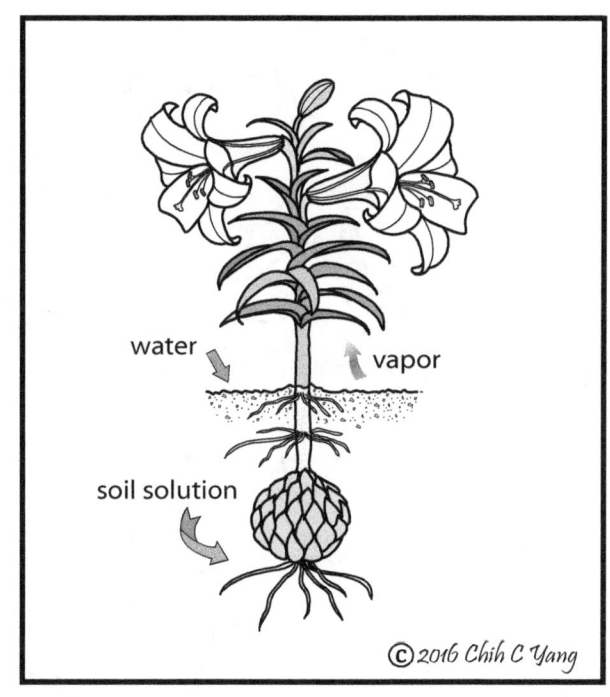

Concrete Thinking
- Fact: Donuts

Abstract Thinking
- Counting: Six

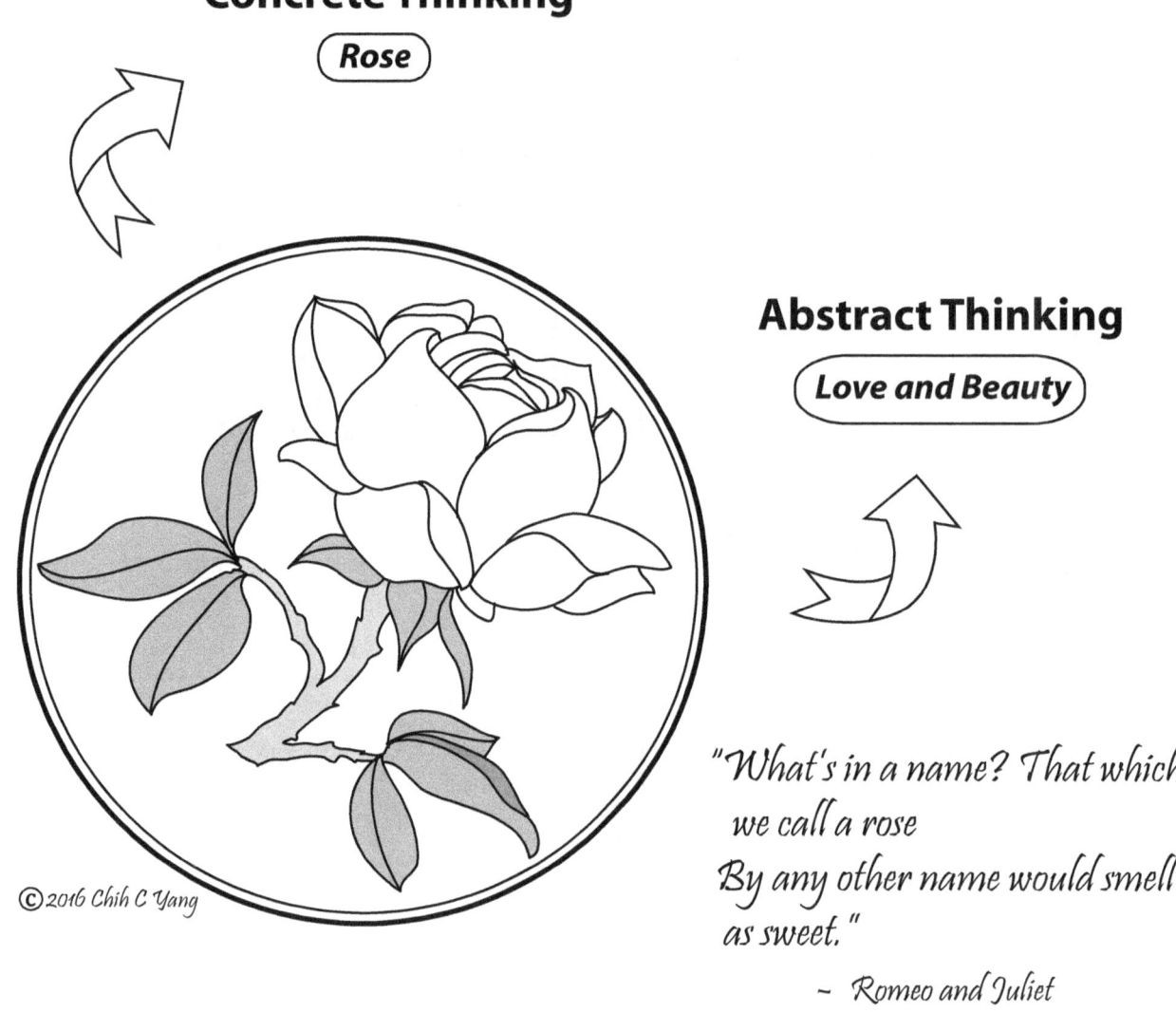

Concrete Thinking (Rose)

Abstract Thinking (Love and Beauty)

"What's in a name? That which we call a rose
By any other name would smell as sweet."

~ Romeo and Juliet
William Shakespeare

What is abstraction?

Abstraction is the process of formulating a **generalized concept** of a common property by disregarding everything else. We capture only the side of facts that are relevant or important in a particular context. We never try to recreate the facts in full.

Abstraction is very natural to human beings that we practice it all the time without being aware of doing so.

Abstraction in Art

Mathematics is the science of patterns; painting is articulation through pattern. (Jensen, 2009)

Modern abstract paintings have evolved from the ***perceptual*** to the ***conceptual***.

The Perceptual

- reproducing an image in full; how an image looks is the most important aspect.

Whistlejacket, George Stubbs, c 1762. (Wikimedia Commons)

The painter, George Stubbs, paid precise attention to the details on the horse.

The Conceptual

- painting an idea or emotion; the idea or emotion behind an image is the most important aspect.

Reiter, Wassily Kandinsky, 1911. (Wikimedia Commons)

Painting was deeply spiritual for the artist, Wassily Kandinsky. He viewed non-objective, abstract art as an ideal visual mode to convey human ideas and emotions.

Blue Horse I, Franz Marc, 1911. (Wikimedia Commons)

In the painting, "Blue Horse", Franz Marc depicts the subject as a sense of harmony, peace and balance.

Abstraction in Patent Law

In patent law, a patent is where an inventor defines the scope of his/her monopoly over an invention. The "invention" refers to a tangible, working apparatus or process.

Example: The Wright Brothers' Patent

The Wright Brothers invented a fully controllable wooden flying machine in 1903.

The machine is a tangible object.

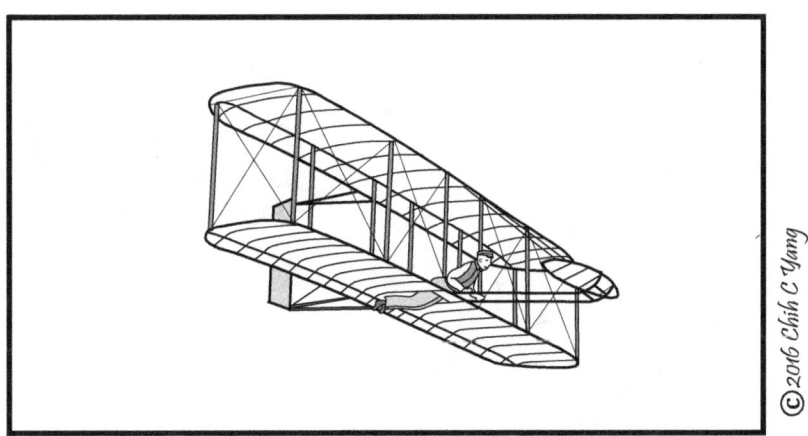

A patent was granted to the Wright Brothers in 1906.

The invention claims of the patent are abstract in form.

> "... In a flying machine, a normally flat aeroplane having lateral marginal portions capable of movement to different positions above or below the normal plane of the body of the aeroplane, such movement being about an axis transverse to the line of flight, whereby said lateral marginal portions may be moved to different angles relatively to the normal plane of the body of the aeroplane, so as to present to the atmosphere different angles of incidence, and means for so moving said lateral marginal portions, substantially as described. ..." (Excerpt - US Patent 831,393)

The Wright Brothers' invention claims refer to the idea and principle of a flying machine. Specifically, it referred to the **definition** of a controllable aircraft, which covered both the wooden machine and most modern aircraft.

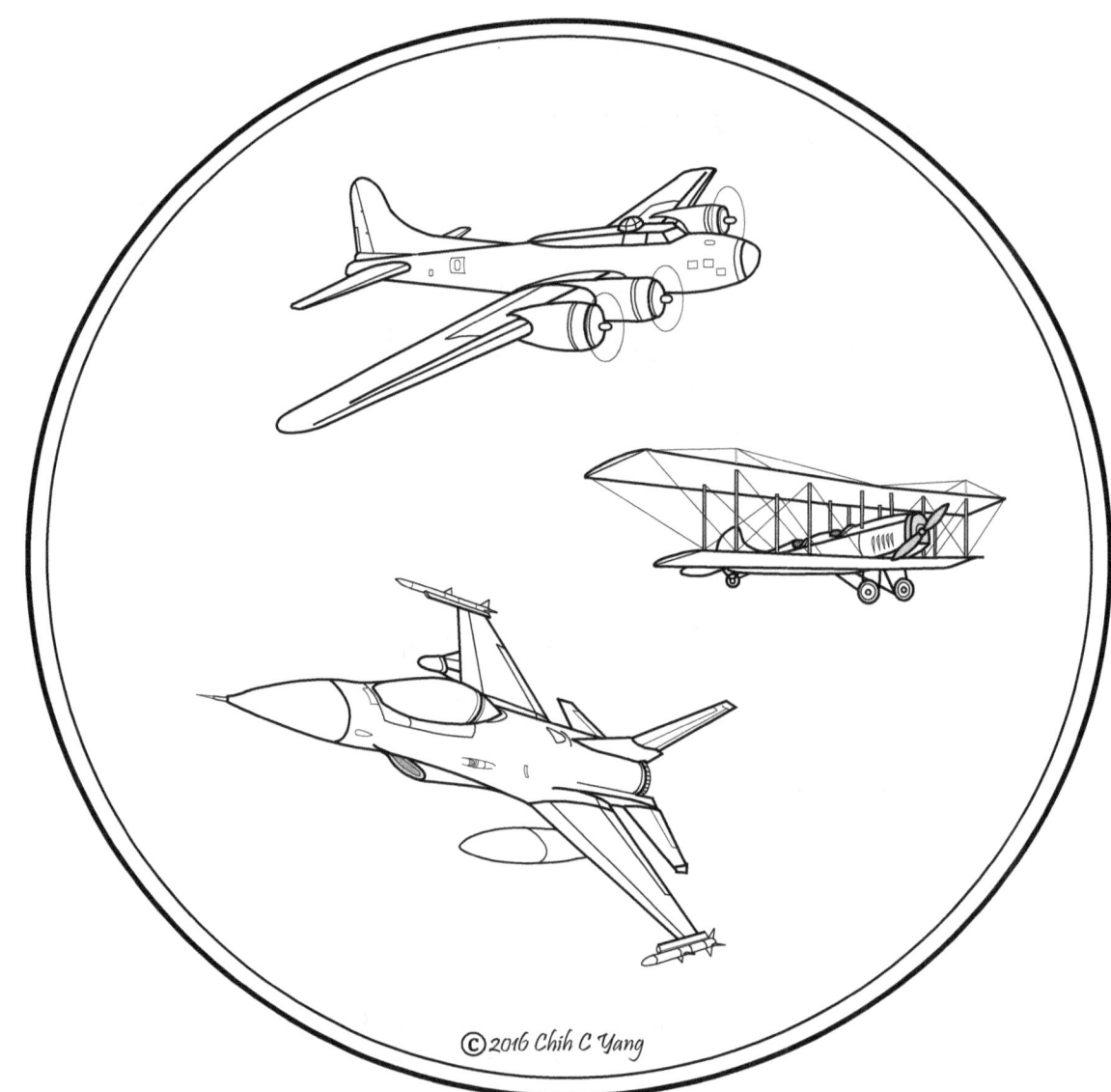

Modern fixed-wing aircrafts still use the same concept of a three-axial control.

Despite over 30 infringement suits, the patent was successfully defended such that nobody was able to break it during the patent's 17 year term.

Abstraction in Computer Science

Abstraction has been a core concept of computer science.

Example: Simplifying a circuit design

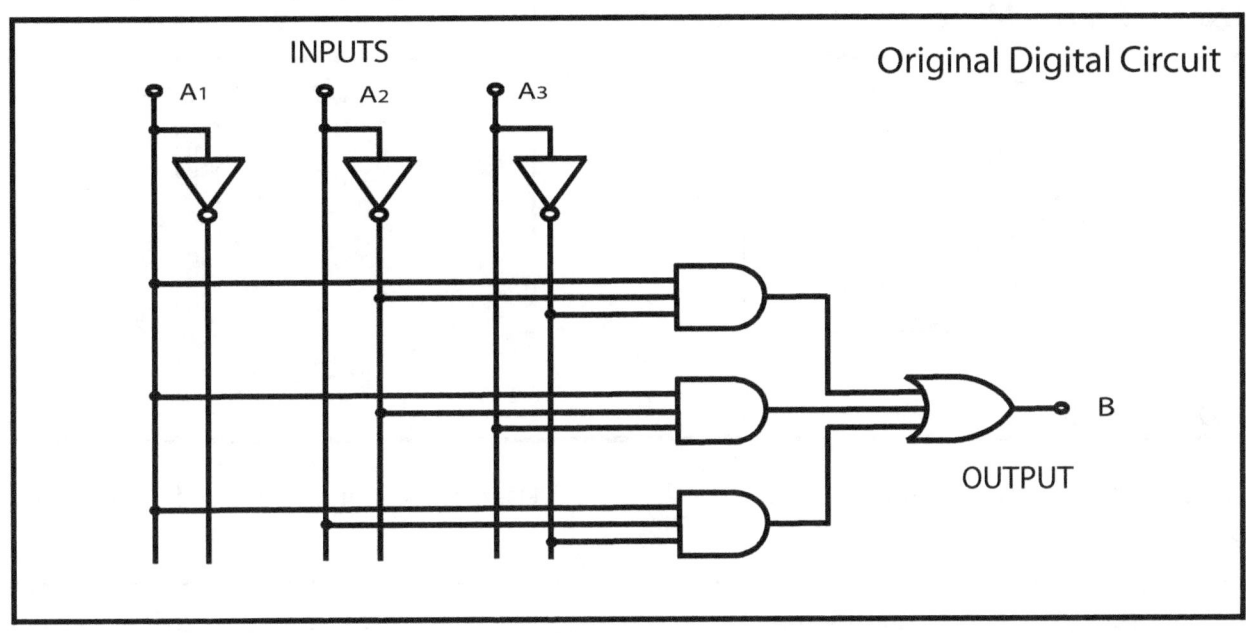

A digital circuit can be expressed as a Boolean function table.

Boolean Function Table

INPUTS			OUTPUT
A_1	A_2	A_3	B
0	0	0	0
0	0	1	0
0	1	0	0
0	1	1	0
1	0	0	1
1	0	1	1
1	1	0	1
1	1	1	0

Abstraction of a circuit

Equivalent

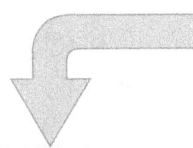

A simpler Boolean expression can be derived from the function table and implemented as a circuit.

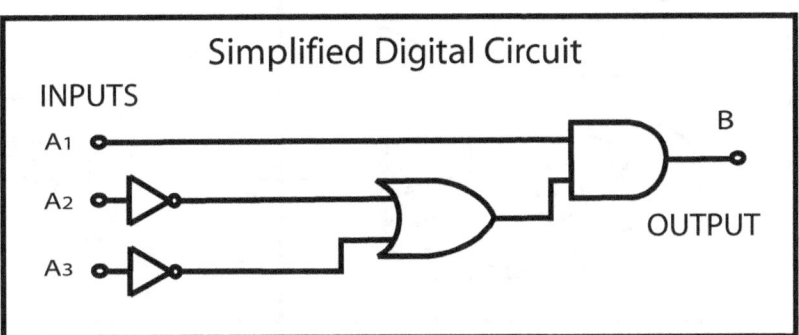

Levels of Abstraction

Performing abstraction in layers reduces the complexity of a system, which in turn leads to efficiency.

Example: The Microwave Oven Design

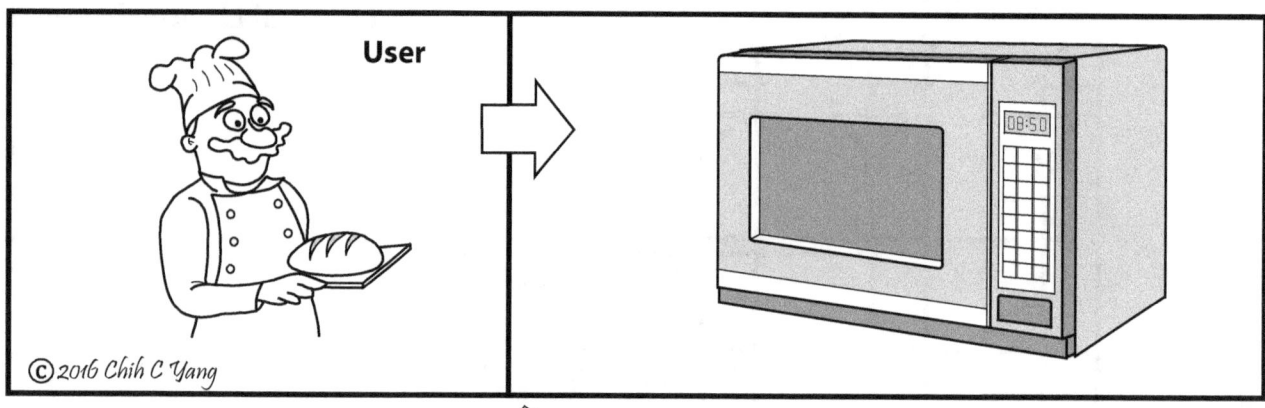

A microwave oven is a "black box" to users.

Users need not be aware of what's inside.

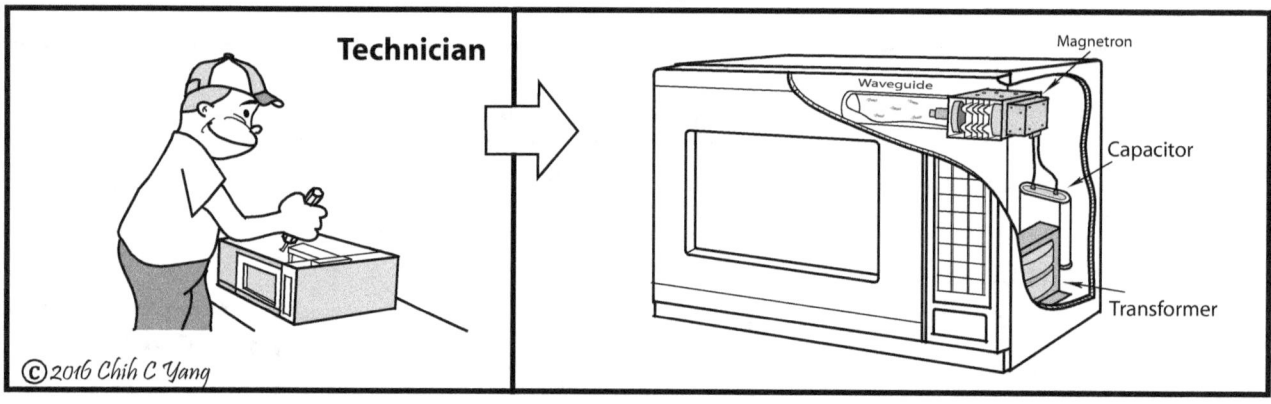

A technician only installs parts but need not know how the parts are made.

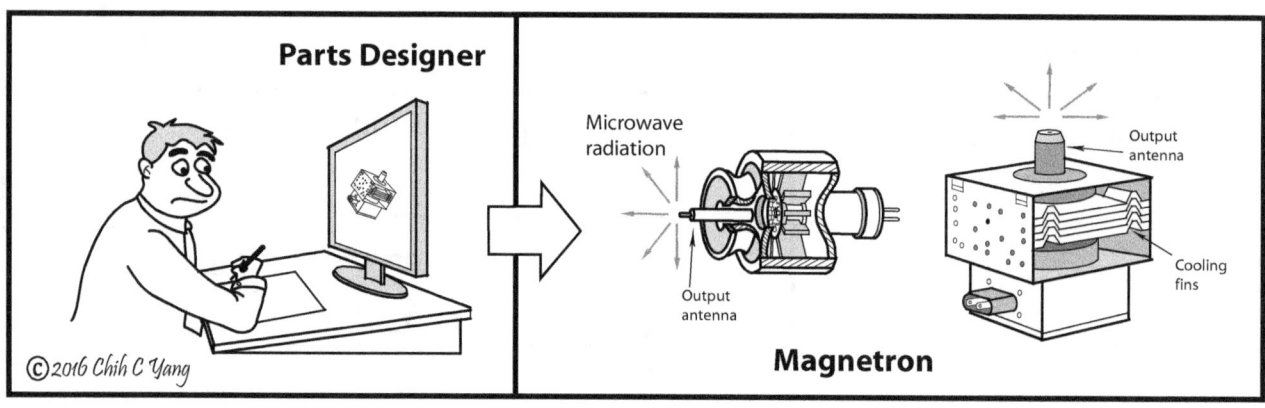

Levels of Abstraction in Computers

The concept of layered abstraction is widely used in computer and information technology.

Example: Computer Architecture

Early computers did not have any form of hardware abstraction.

In the early days, programmers had to operate the machines in person.

Operating a machine was complicated, difficult and tedious.

Creating multi-levels of abstraction drastically simplified computing systems.

With multi-levels of abstraction, programmers only needed to understand and focus on the level they were working on.

- **Level 4:** Operating System (Windows, Linux, Google Android)
- **Level 3:** Virtual machine interpretation
- **Level 2:** Microprogramming
- **Level 1:** Digital Logic
- **Level 0:** Electronic circuit

© 2016 Chih C Yang

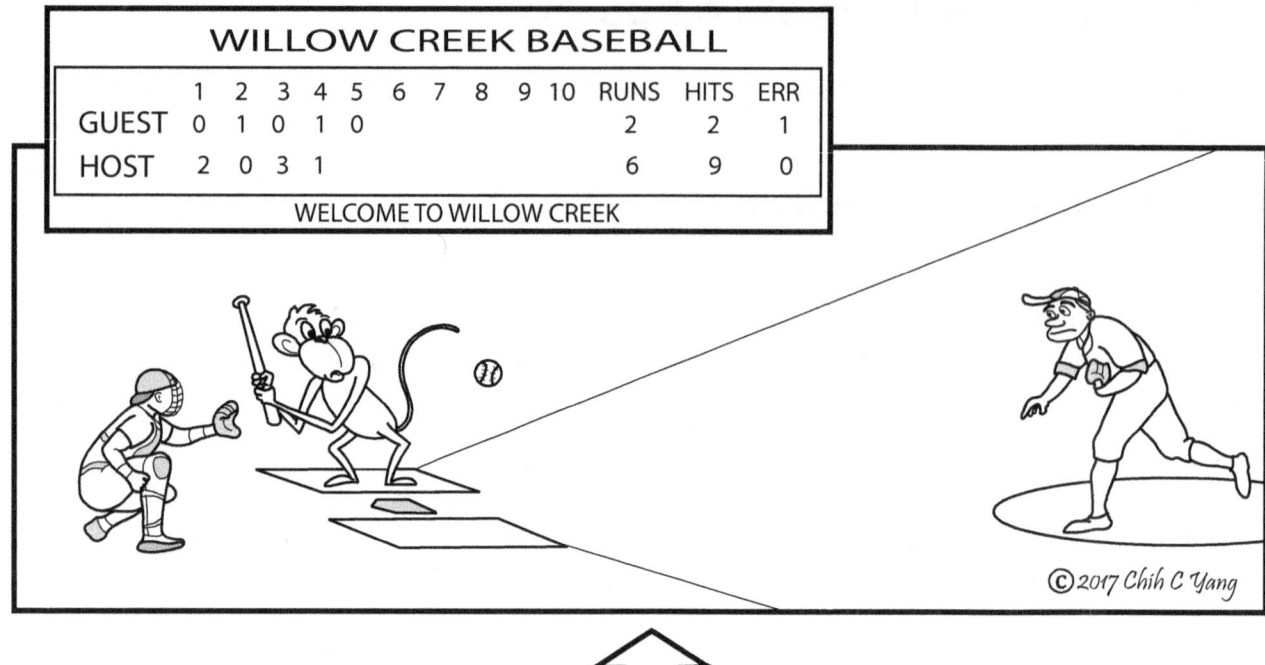

Abstract thinkers may recognize that strategies in baseball are like marketing products.

Abstract thinkers can perceive analogies and relationships that others may not.

Application: The Model-T Ford Assembly Line

In 1913, Ford Motor Company manager, William C. Klann, visited a Chicago slaughterhouse. There, animal carcasses moved by trolleys allowing each butcher to perform a specialized task. This efficient model of cutting up a carcass into cuts of meat gave Klann an idea.

"If they can go from a slaughtered animal to cuts of meat in this step-by-step process, why can't we assemble cars in the same way?"

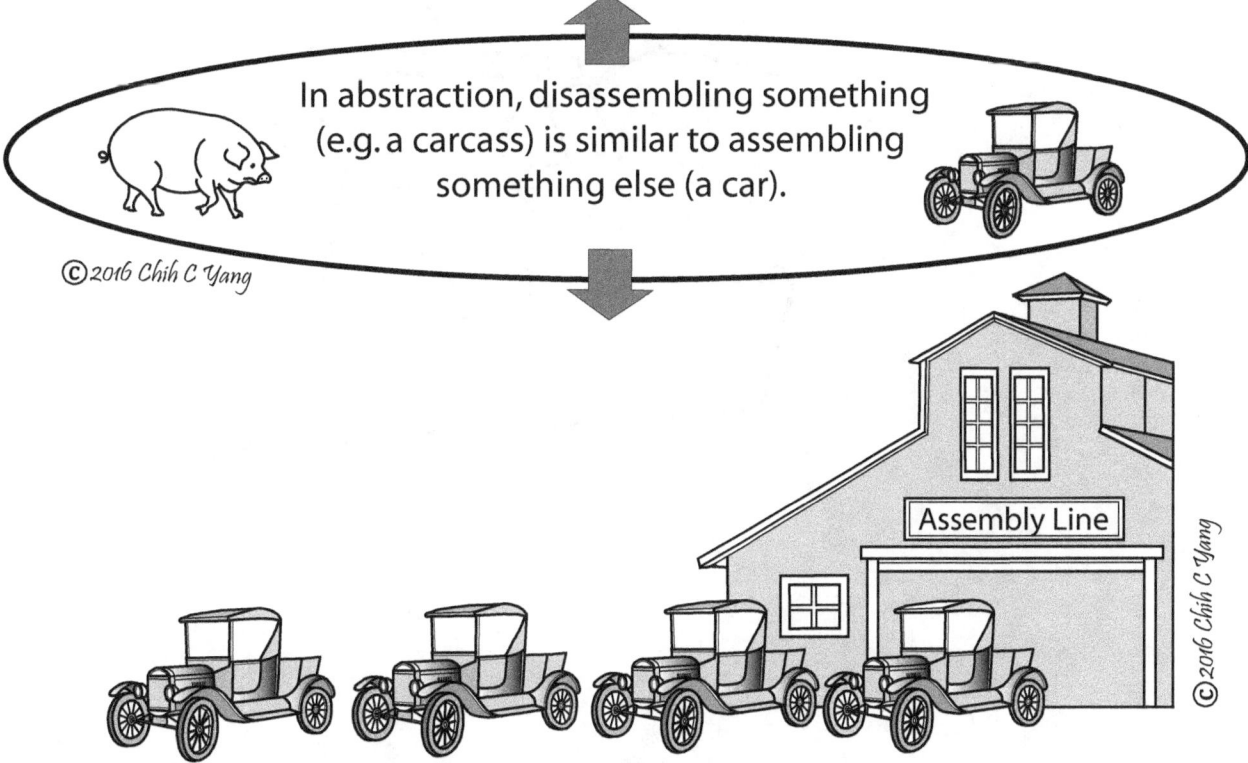

On December 1, 1913, Ford Motor Company introduced the world's first moving assembly line and revolutionized the automobile industry.

**Abstraction allows you to see things differently.
It helps you to think outside the box
as a way to find creative solutions to problems.**

Thought Experiments

A thought experiment is a way of thinking abstractly. It allows us to test hypotheses or theories in our imagination rather than in a laboratory. Thought experiments are invaluable whenever it would be impractical or even unethical to perform a real test. Oftentimes, thought experiments can help us to clarify and understand abstract concepts or situations.

Example - Plato's Allegory of the Cave

In Book VII of Plato's *Republic*, Plato uses the allegory of men imprisoned in a cave trying to make sense of the real world. He uses this allegory to help readers see that we rarely, if ever, see reality directly. Instead, we see it through the filter of what we already think we know.

Plato's Allegory of the Cave (continued)

Imagine prisoners who have been chained inside a cave since childhood. In the cave, a fire blazes behind the captors casting shadows. From the prisoners' perspective, all they can see are the shadows of men passing by carrying objects such as statues or figures of animals. Though these shadows are all the prisoners knew, they can make deductions about what life must be like in the outside world.

If the prisoners were to be released, what would they think when they saw the reality that the shadows represent? According to Plato, "to them, I said, the truth would be literally nothing but the shadows of the images."

Plato's allegory of the cave illustrates that people may feel enlightened about something even though they are still in the dark about how it actually works. The allegory of the cave demonstrates how exiting the cave can lead to a clearer understanding of the world.

There have been many different interpretations of this allegory, which has often been used as a metaphor by many Western writers.

A modern prisoner of the cave

The shadows in Plato's metaphor are similar to our electronic devices. We see the real world through cyber-virtuality.

Little Johnny's Thought Experiment - Traveling

Little Johnny likes to travel around the world. This is how Johnny imagines a trip to China would be using a thought experiment.

1. He travels in a hot air balloon, which rises high above the ground.
2. The balloon then floats in place in the air, and does not move with the earth's rotation.
3. When his destination is reached, Johnny descends in the balloon.

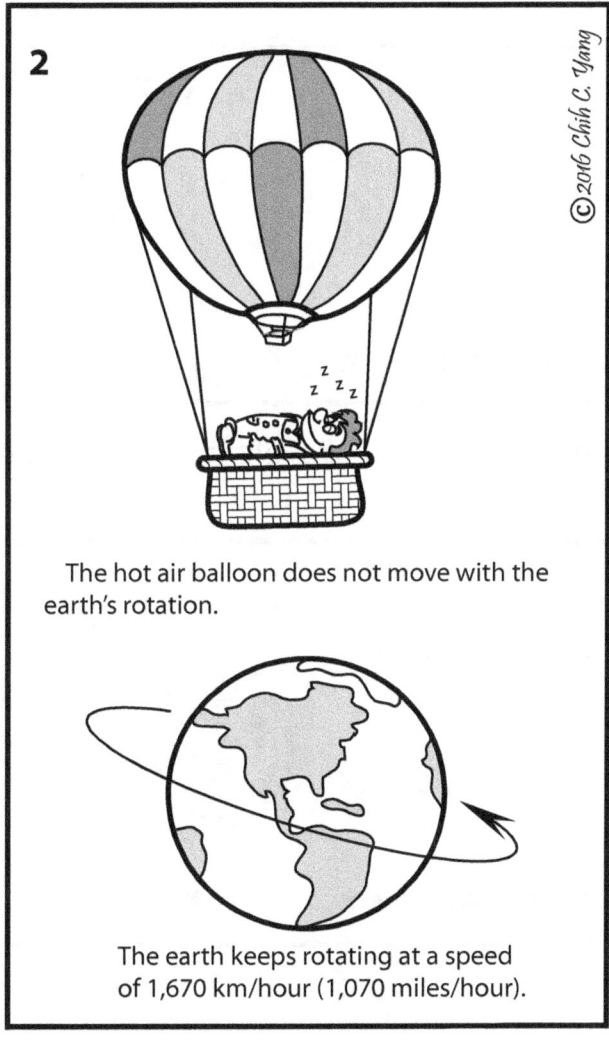

The hot air balloon does not move with the earth's rotation.

The earth keeps rotating at a speed of 1,670 km/hour (1,070 miles/hour).

Was Johnny's thought experiment correct? Why or Why not?
(Please see Galileo's sailing-ship thought experiment)

The Moving Earth

Until the 16th century CE, most scientific research was based on natural observations or thought experiments. Aristotle and his followers never performed physical experiments.

In the same era, Copernicus and others demonstrated that heavenly bodies orbit around the sun. They introduced the notion that the earth spins on its axis, making one rotation every 24 hours. At first, there was strong opposition to these ideas, and not just from religious authorities.

In Defense of the Earth's Immobility

The Aristotelian view

Aristotelian scientists believed that the earth did not move. According to this view, if the earth spun on its axis, a cannon ball fired eastward would not advance far from the cannon because the rotating earth would carry the cannon swiftly eastward.

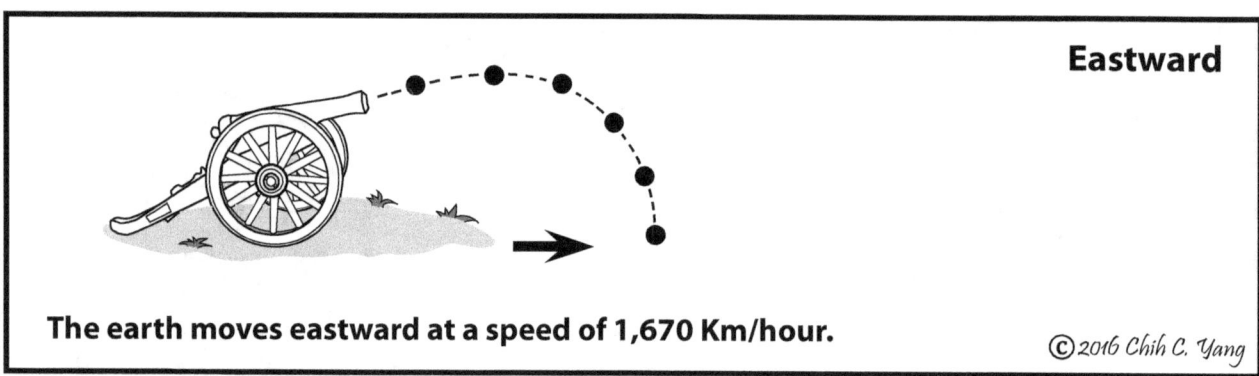

The earth moves eastward at a speed of 1,670 Km/hour.

By the same reasoning, a cannon firing westward should travel farther from the cannon.

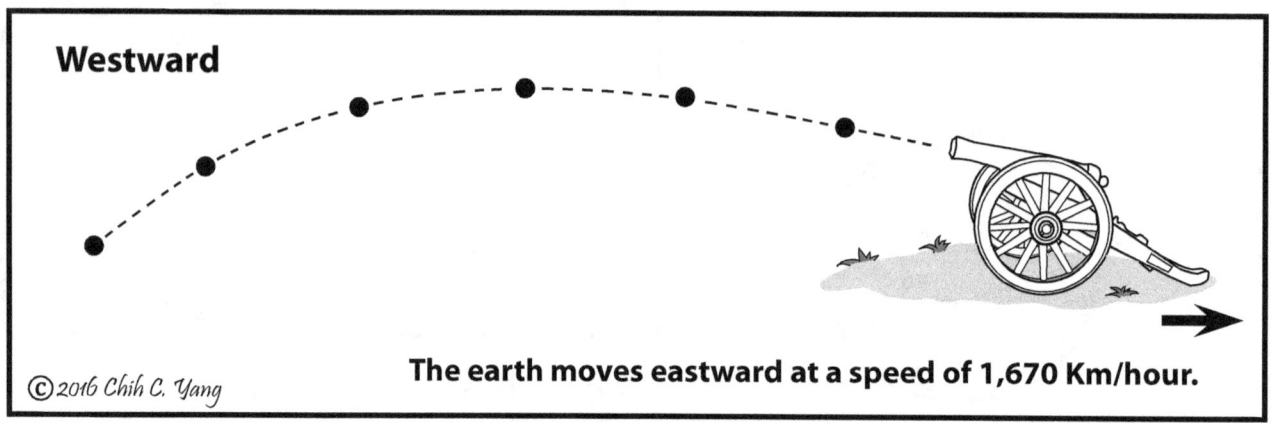

The earth moves eastward at a speed of 1,670 Km/hour.

Observations showed no such discrepancies. A cannonball travels equal distances fired in either direction.

Galileo's Sailing Ship

In response to the Aristotelian scholars, Galileo Galilei illustrated what would happen if a cannonball were dropped from the mast of a moving ship. In this thought experiment, the moving ship is analogous to the moving earth.

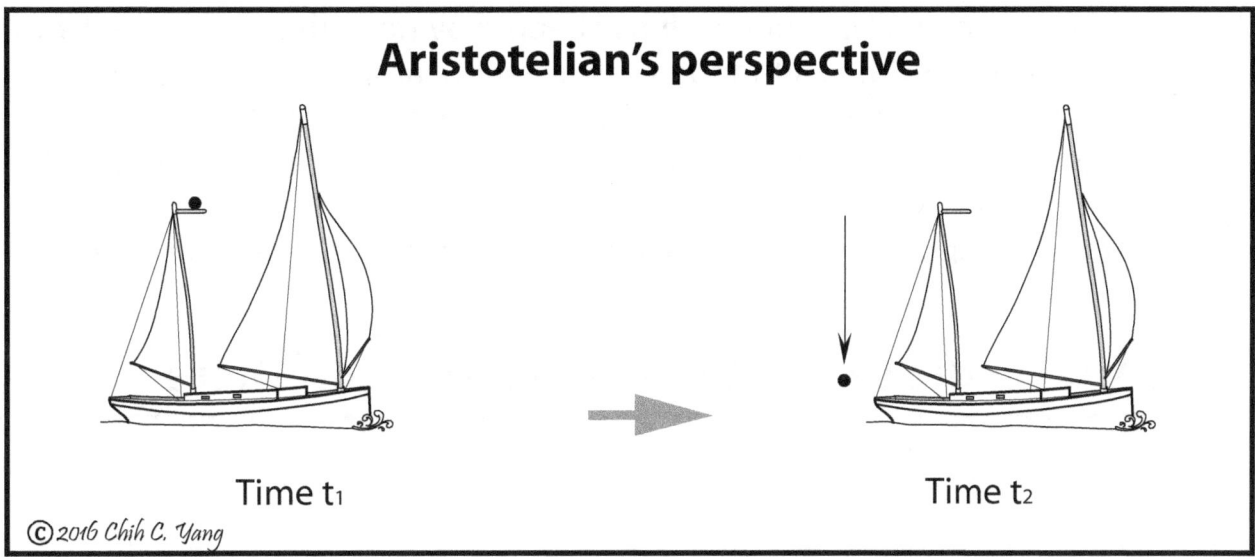

According to the Aristotelian view, a sailing ship would move faster than the cannonball and would therefore leave the falling ball behind. The ball would thus land behind the mast or behind the ship.

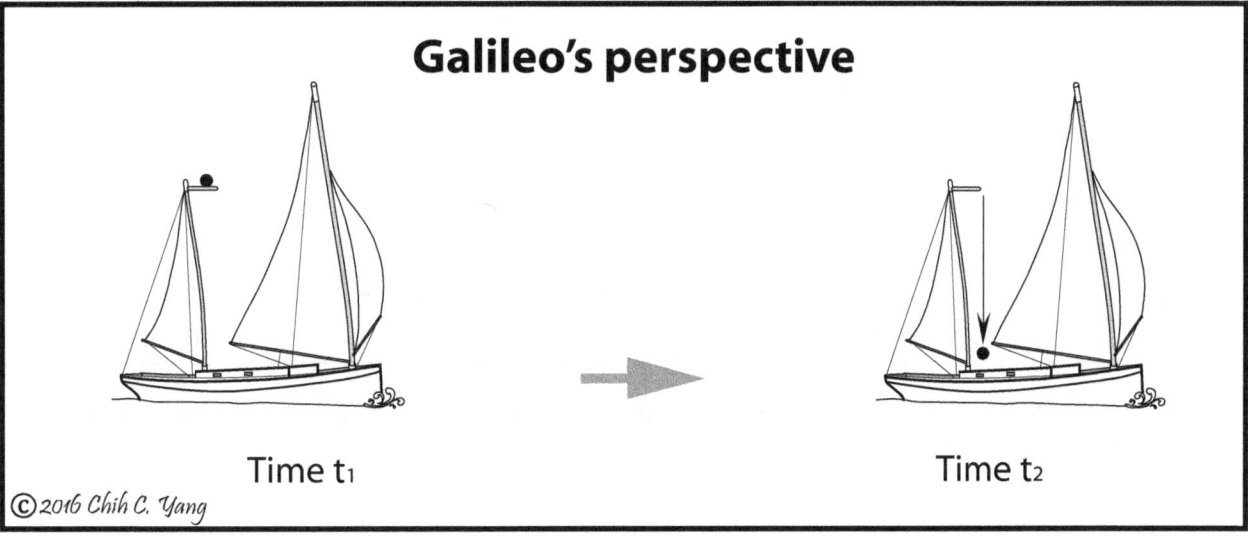

From Galileo's perspective, an observer on the ship would see the ball dropped from the top of the mast fall and hit the base of the mast. Performing this experiment showed that it is possible that the earth is moving.

Galileo explained this phenomenon in terms of **the law of Inertia**, a law based on his experiments using an inclined plane.

Galileo's Sailing Ship (continued)

> **The Law of Inertia** states that a moving object tends to continue in the same speed and direction until acted upon by an outside force.

Before the cannon ball is dropped, it has been moving at the same speed and direction as the sailing ship. Although the ball's vertical speed changes, its horizontal speed remains constant to the ship's. Crew members on the ship, who have the same horizontal inertia as the ship and ball, would see the dropped ball fall straight down.

However, a bystander on the shore who would see the ball falling in a different trajectory.

Observing the Sailing Ship from a Point on Shore

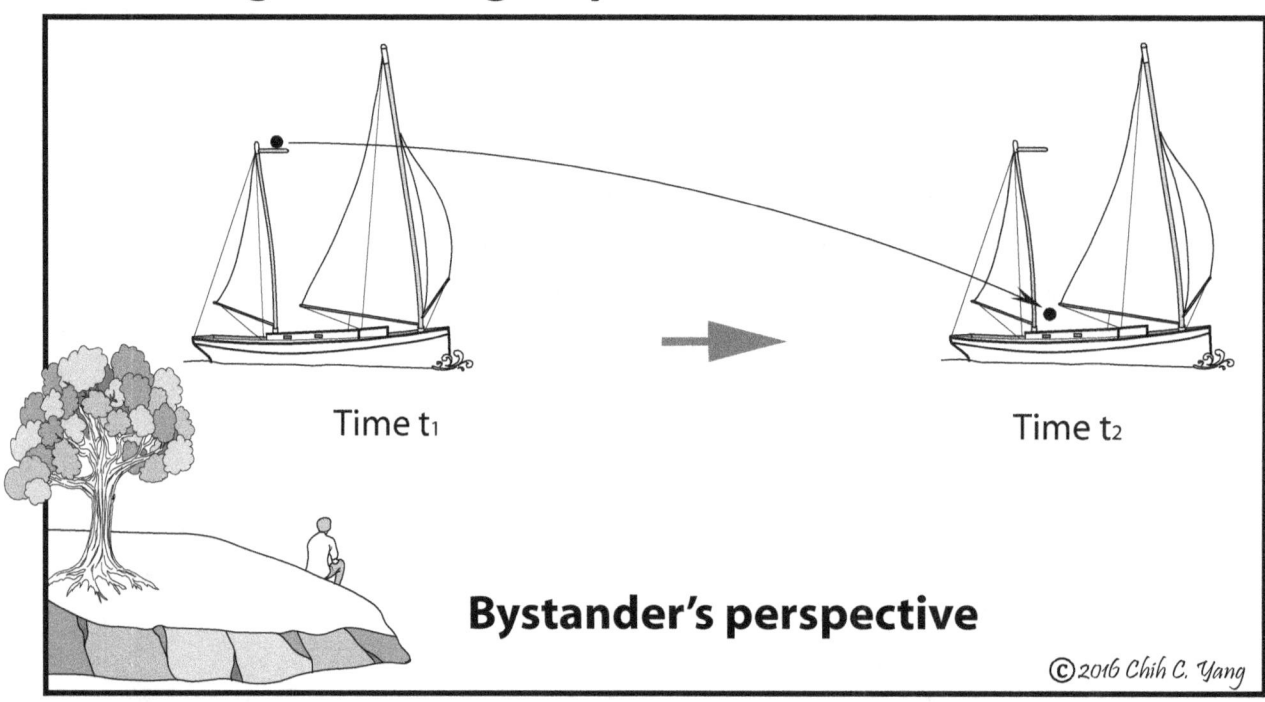

Time t₁   Time t₂

Bystander's perspective

©2016 Chih C. Yang

A bystander on shore sees the ball continue to move horizontally as it falls. In this case, the ball's trajectory is parabolic.

Galilean Relativity

Galileo discovered that all motion is relative. When a person on a ship watches another ship pass by, he/she cannot tell which ship is moving without referring to some other object, like a tree or the shoreline. Centuries later, the concept of relativity became the basis of Einstein's Theory of Special Relativity.

Einstein's Train

Albert Einstein is the master of thought experiments. He created a great show of thought experiments to explain *relativity*.

Example - Time Dilation

According to Einstein's Theory of Special Relativity, time slows as one approaches the speed of light. A thought experiment is conducted to explain this phenomenon.

In the thought experiment, Einstein is riding on a fast moving train with a light clock.

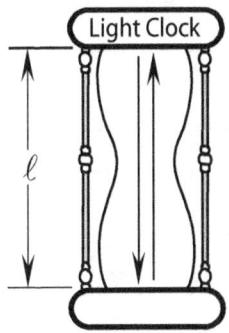

A light clock consists of two mirrors that bounce a light pulse back and forth. One bounce of light from one mirror to the other and back counts as one tick of the clock.

Now imagine Einstein and the light clock moving at the same speed as the train. From Einstein's perspective, the light pulse travels straight up and down between the mirrors.

Einstein's Train (continued)

To a bystander, however, the light pulse in the clock does not go straight up and down. Instead, it zigzags up and down in a diagonal path. Thus, the light pulse on the moving train has a longer distance to travel. So, Einstein's light clock on the train appears to be ticking slower than the light clock from the bystander's point of view.

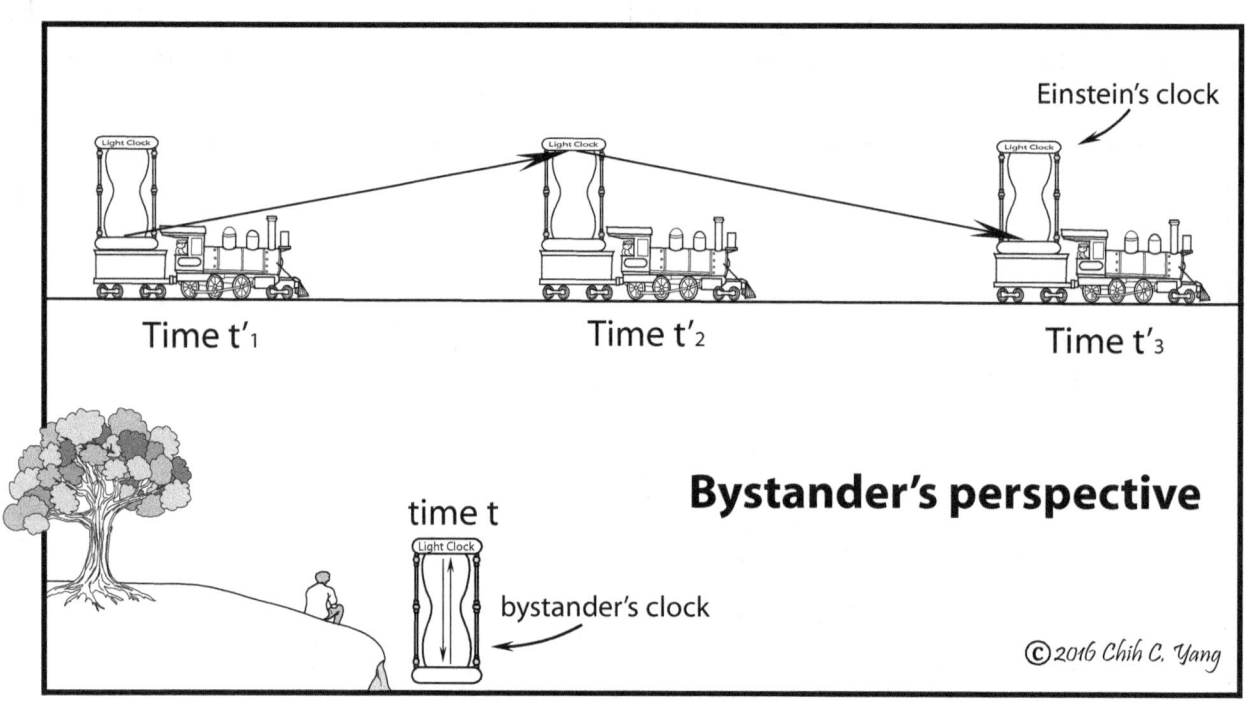

Let, ℓ = height of the light clocks,
v = Einstein' speed
t' = time on Einstein's clock
t = time on the bystander's clock
c = speed of light

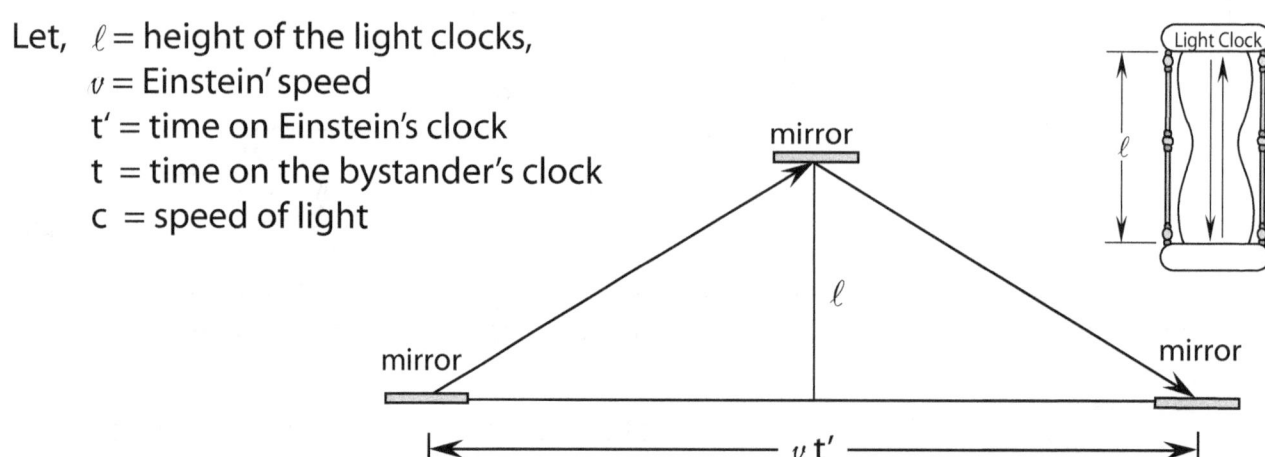

Using algebra, we can create a formula to calculate how much time slows down in the train.

When v is close to the speed of light (c), t' becomes greater than t. This means that Einstein's time slows down as the train approaches the speed of light.

$$t' = \frac{t}{\sqrt{1 - \frac{v^2}{c^2}}}$$

Einstein's Twin
- A Thought Experiment in Special Relativity

Al and Bert are identical twins. While Al remains on Earth, Bert makes a journey into space in a rocket at a speed close to that of light.

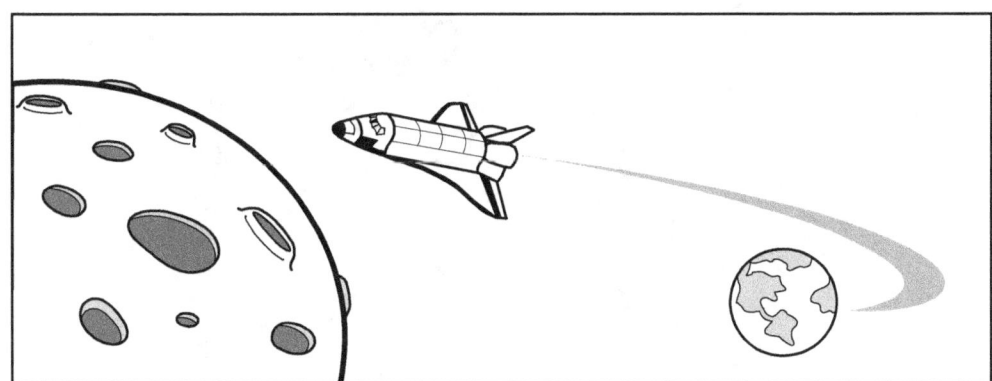

Due to *time dilation* (see previous section), Bert's time slows down when he travels at very high speed.

Later, Bert returns home to find his brother has aged.

Thought Experiment Applications

Thought experiments are commonly used in physics and philosophy. They are also useful in mathematics, biology, computer science, economics, law and finance.

Example - The Infinite Monkey Theorem in Mathematics

In mathematics, the Infinite Monkey Theorem states that a monkey striking keys randomly on a keyboard for an infinite amount of time would eventually type meaningful text, such as William Shakespeare's sonnets.

The thought experiment is used to illustrate the nature of probability over an infinite period of time.

Though the monkey's chances of typing a Shakespeare's sonnet is incredibly small, it is not equal to zero.

From Thought Experiments to Computer Experiments and on to Virtual Reality

With advances in technology, the relationship between thought experiments and computer experiments has attracted the attention of scientists and philosophers. It has been argued that computer experiments will play an increasingly significant role in the future, especially in the fields of complex phenomena. Some scientists have proposed using video games as executable thought experiments.

Virtual Reality

Virtual reality is a computer-generated 3-D artificial environment that simulates a real or imagined world. Users can interact with this artificial environment using special electronic devices.

References – Abstraction and Thought Experiments

[1] Bajnok, Belo; An Invitation to Abstract Mathematics, Springer, 2013

[2] Gamwell, Lynn; Mathematics and Art: A Cultural History, Princeton University Press, NJ, 2016

[3] Gray, Jeremy; Plato's Ghost: The Modernist Transformation of Mathematics, Princeton University Press, 2008

[4] Heath, Thomas Little, Sir; A History of Greek Mathematics, The Clarendon Press, 1921

[5] Kleiner, Israel; A history of Abstract Algebra, Birkhauser Boston, 2007

[6] Maddox, Randall B.; A Transition to Abstract Mathematics – Mathematical Thinking and Writing, 2nd Edition. Elsevier, UK, 2009

[7] Pinter, Charles C.; A Book of Abstract Algebra, 2nd Edition. McGraw-Hill Publishing Company, 1990

[8] Plato and Benjamin Jowett; The Republic from an 1892 edition, The floating Press, 2009

[9] Pohlen, Jerome; For Kids Series: Albert Einstein and Relativity for Kids: His Life and Ideas with 21 Activities and Thought Experiments, Chicago Review Press, 2012

[10] Siu, Man-Keung; Why Is It Difficult to Teach Abstract Algebra?, The University of Hong Kong, 2004

[11] Sorensen, Roy A.; Thoughts Experiments; Oxford University Press, New York, 1998

[12] Stallings, William; Computer Organization and Architecture, 7th Edition, Pearson Prentice Hall, NJ, 2006

[13] Wright, Orville and Wilbur Wright; Flying Machine, US Patent No. 821,393, 1903

[14] Webb, Stephen; Out of This World: Colliding Universe, Branes, Strings, and Other Wild Ideas of Modern Physics, 2004, Copernicus Books, 2004

4

Set and Infinity

Why Set Theory?

Know your foundation before build a house

Set theory is the foundation of modern mathematics.

Set Theory

Most disciplines of mathematics have arisen from the cumulative efforts of many mathematicians, over multiple generations.

But in the late 19th century, set theory was created by one individual - Georg Cantor (1845 -1918).

Infinite Sets

It was Cantor who taught us how to compare the magnitude of infinite sets by using the same methods to compare finite sets -- through clearly defined concepts such as cardinality and denumerability.

Zeno's Paradox: Achilles and the Tortoise

Before Cantor's time, the infinite was poorly understood. The infinite in mathematics was obscure and replete with paradoxes (contradictions). One such example is the Paradox of Zeno's (490 - 425 BC).

© 2016 Chih C Yang

In a race, Achilles gives the tortoise a head start of, let's say, 10 meters. Achilles has to run 10m to reach the starting point of the tortoise, but when he reached that point, the tortoise has advanced farther. Whenever Achilles reaches any point along the tortoise's rout, he still has farther to go. Because there are infinite number of points Achilles must reach along the tortoise's path, he can never overtake the tortoise.

During the Middle Ages, mathematics in Europe almost came to a standstill. The concept of infinity had become a theological subject rather than a scientific one.

Most models of the universe assumed it to be finite. The only infinity was God.

It was not until the sixteenth century that infinity was accepted and written about as a legitimate subject of inquiry in mathematics.

How Big is Infinity?

Counting in the Finite World

Before the numbers were invented, shepherds used to keep a pile of pebbles with one pebble for each sheep. When the shepherd matched the pebbles one to one with his sheep and had pebbles left over, then, some sheep were missing.

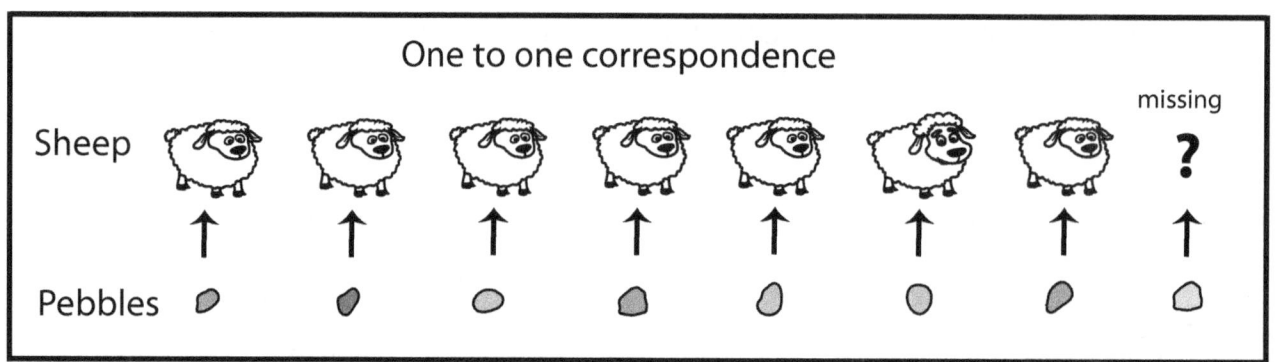

After the numbers were invented, it was possible to count things. In **counting**, pebbles were replaced by the natural numbers in one-to-one matching.

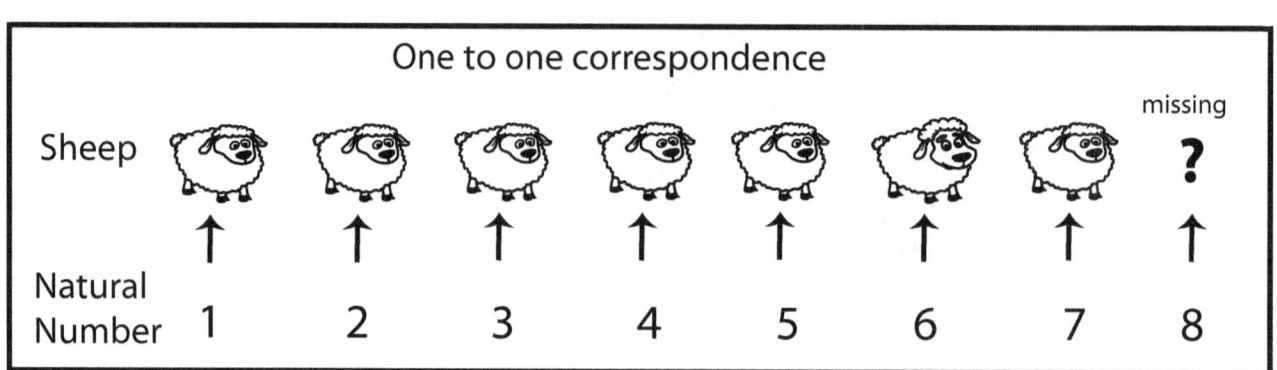

Counting - a more abstract form of one-to-one matching than using pebbles -- works well in comparing the size of finite sets. But when mathematicians are working with infinite sets, comparing by counting can lead to paradoxes.

Counting an Infinity Set

Paradox 1

Imagine trying to count all of the even positive integers by one to one correspondence.

Even Numbers	2	4	6	8	10	12	14	...	$2n$	...
	↑	↑	↑	↑	↑	↑	↑		↑	
Natural Numbers	1	2	3	4	5	6	7	...	n	...

From the chart above, you can see that for every even number there is a corresponding natural number. Thus, it looks as though the set of even numbers and the set of all natural numbers (including the odd numbers) are "the same size."

Paradox 2

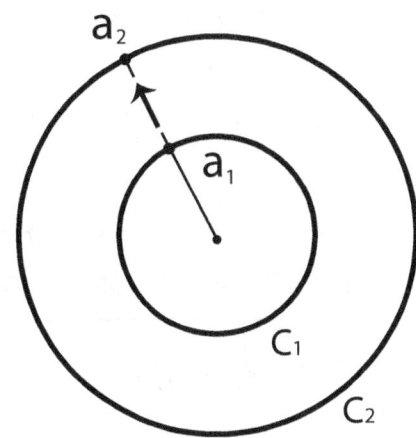

Another example of comparing size by one-to-one matching:
C_1 and C_2 are two concentric circles, and point a_1 and point a_2 lie on the same radial line from the common center of the circles. For every radial line we draw, we can establish a one-to-one correspondence between a point on C_1 and a point on C_2.

Thus, we might conclude that circles C_1 and C_2 are the "same size."

A paradox is a statement that seems contradictory or illogical but that may still be true.

Comparing Infinite Sets

We need precise terms to describe what is being compared.

Subset and Cardinality

You are familiar with the set of natural numbers, **N**, and the set of even natural numbers, **E**. Every element of **E** is also an element of **N**. Thus, we say that **E** is a **subset** of **N**, $E \subset N$.

The number of elements in set A is called **cardinality**, denoted by |A|.

Cantor distinguished infinite sets by their "size." If two sets could be put into a one-to-one correspondence, then these two sets were considered to be the "same size." That is, the two sets were numerically equivalent or equal cardinality.

There is a one-to-one correspondence between the set of natural numbers and the set of even natural numbers.

```
Natural numbers:           1   2   3   4   5   ...
one-to-one correspondence  ↕   ↕   ↕   ↕   ↕         |N| = |E|
Even natural numbers:      2   4   6   8   10  ...
```

Thus, the set of natural numbers and the set of even natural numbers are numerically equivalent.

Countable Infinite Sets

The cardinality of the set **N** of natural numbers is called $\aleph_0$ (aleph zero). Cantor envisioned a whole system of infinite cardinalities $\aleph_0, \aleph_1, \aleph_2, \ldots$, of which $\aleph_0$ is the smallest. Any set that is numerically equivalent to **N** is called a countable, or denumerable infinite set.

The set of integers, **Z**, is countable. $|Z| = |N| = \aleph_0$

Integers, **Z**	0	1	-1	2	-2	3	-3	4	-4	...
	↑	↑	↑	↑	↑	↑	↑	↑	↑	
Natural Numbers, **N**	1	2	3	4	5	6	7	8	9	...

The square of natural numbers, n^2, is countable.
$$|N^2| = |N| = \aleph_0$$

The square, n^2	1	4	9	16	25	36	49	...	n^2	...
	↑	↑	↑	↑	↑	↑	↑		↑	
Natural Numbers, n	1	2	3	4	5	6	7	...	n	...

The set of odd natural numbers, **A**, is countable.
$$|A| = |N| = \aleph_0$$

Odd numbers	1	3	5	7	9	11	...
	↑	↑	↑	↑	↑	↑	
Natural numbers	1	2	3	4	5	6	...

The union of two countable sets is countable. For example, the sets of even natural numbers **E** and odd natural numbers **A** are both countable.

The union of **E** and **A**, **E** ∪ **A** = **N**

Since $|E| = |A| = |N| = \aleph_0$, $|E \cup A| = \aleph_0 + \aleph_0 = |N| = \aleph_0$ Thus, **E** ∪ **A** is countable

Hilbert's Hotel

Hilbert's Hotel is a thought experiment which illustrates an interesting property of infinite sets.

Whenever a new guest arrives, the manager frees up room 1 for the newcomer by shifting the current occupant from room 1 to room 2, room 2 to room 3, and so on. By repeating this procedure, it is possible to make room for any finite number of new guests.

The cardinality of hotel's occupants plus one is $\aleph_0 + 1 = \aleph_0$.

When there are infinitely many new guests arrive, the manager moves the occupant of room 1 to room 2, room 2 to room 4, room 3 to room 6, and, in general, room n to room 2n. This opens up all the odd-numbered rooms (infinitely many of them) for the newcomers.

The cardinality of hotel's current occupants plus that of infinitely new guests is $\aleph_0 + \aleph_0 = \aleph_0$.

The cardinality of infinite sets is not a number, does not obey the rules of arithmetic.

Is the Set of Rational Numbers Countable?

Arrange the rational numbers p/q (q ≠ 0) in a square array:

⋯	4/-3	4/-2	4/-1	4/1	4/2	4/3	4/4	⋯
⋯	3/-3	3/-2	3/-1	3/1	3/2	3/3	3/4	⋯
⋯	2/-3	2/-2	2/-1	2/1	2/2	2/3	2/4	⋯
⋯	1/-3	1/-2	1/-1	1/1	1/2	1/3	1/4	⋯
⋯	-1/-3	-1/-2	-1/-1	-1/1	-1/2	-1/3	-1/4	⋯
⋯	-2/-3	-2/-2	-2/-1	-2/1	-2/2	-2/3	-2/4	⋯
⋯	-3/-3	-3/-2	-3/-1	-3/1	-3/2	-3/3	-3/4	⋯

We define a mapping between the natural numbers, **N**, and rational numbers, **Q**, by following the path:

[Diagram: the same square array with a spiral path starting at 1/1, going to 1/2, then spiraling outward through the array.]

There are repetitions like 2/2 = 3/3 = 4/4 …, 1/2 = 2/4…, etc. Eliminate these, we got one to one correspondence.

Rationals **Q**	1	1/2	2	-2	-1	-1/2	-1/3	1/3	…
	↑	↑	↑	↑	↑	↑	↑	↑	
Natural **N**	1	2	3	4	5	6	7	8	…

$|Q| = |N| = \aleph_0$, **Q** has the same cardinality as **N**. Hence the set of rational numbers is countable.

Real Numbers

The real numbers consist of all the rational and irrational numbers. The irrational numbers were not completely understood before the 19th century. If we consider the representation of numbers as points along a straight number line, the rational numbers are not sufficient enough to cover the entire line. They are **dense**, but leave infinitely many "holes." However, the set of real numbers, in which all holes along the number line are filled, has the property of "**completeness**."

The Set of Real Numbers is Not Countable

There is no one-to-one correspondence between the natural numbers **N** and the real numbers **R**. Cantor called the cardinality of real numbers **c**, $|R| = c$, and demonstrated that the cardinality of the real numbers is greater than $\aleph_0$.

Calculus and Real Numbers

In the 17th century, calculus was developed by two mathematicians working independently - Isaac Newton (1642-1727) of England and Gottfried Wilhelm Leibniz (1646-1716) of Hanover (part of present-day Germany).

A Problem on the Foundation of Calculus

Later in the 19th century, mathematicians realized that to "rigorously" prove the theorems in calculus required the continuous line of real numbers, not the rational numbers, a line with many "holes." This problem was solved by Richard Dedekind in 1872.

Plugging the Holes on the Number Line Took Nearly 100 Years.

Dedekind's Cut demonstrates holes in the rational number line.

Example:
What happens when we try to cut the rational number line at the location of an irrational number like $\sqrt{2}$?

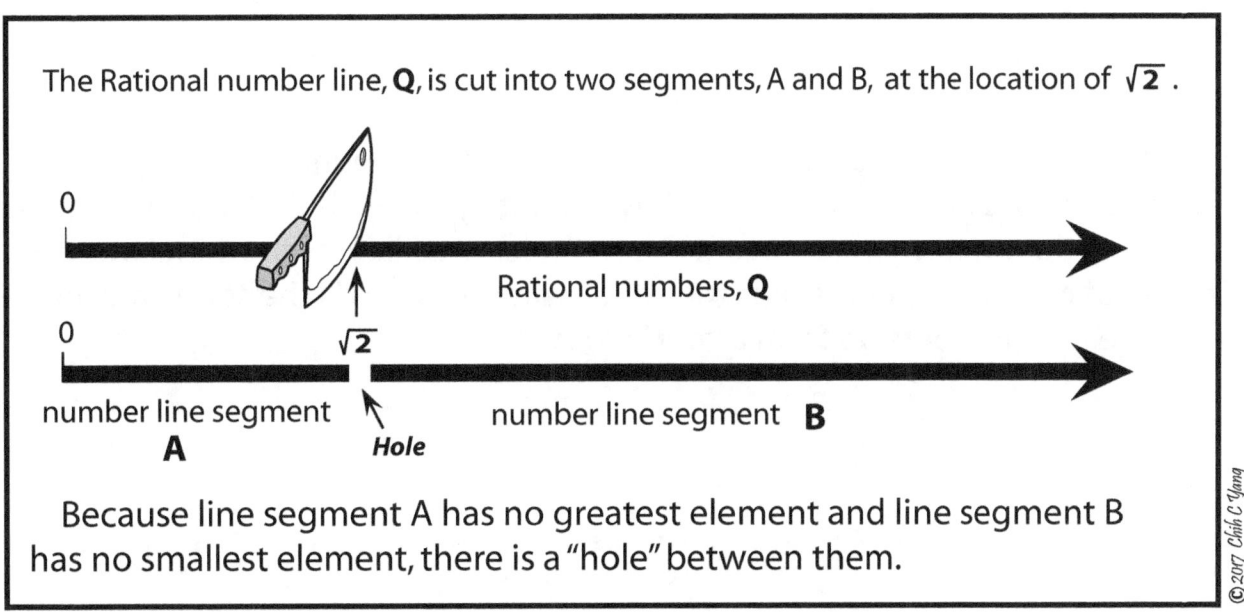

Because line segment A has no greatest element and line segment B has no smallest element, there is a "hole" between them.

If line segment A has a greatest element or line segment B contains a smallest element, the cut defines a rational number, (i.e., no hole).

Richard Dedekind (1831-1916) published his essay *Continuity and Irrational Numbers* in 1872. Rational numbers have infinitely many "holes" in the number line. They do not form a *continuum*. These holes are filled by irrational numbers. The set of rational numbers plus the set of irrationals yields the set of real numbers. The real numbers do form a *continuum*.

Paradox in Set Theory

In 1901, Bertrand Russell (1872 - 1970) demonstrated the following paradox in Set Theory:

> Let **R** be a set
> $$R = \{x \mid x \text{ is a set that does not belong to itself}\}$$
>
> Is **R** a member of itself?

If R is a member of itself, then one member of the set that does not fit the definition.

If R is not a member of itself, then the set R does not include all sets that are not members of themselves.

It shows that not every collection can be considered as a set.

Russell's paradox caused mathematicians to rethink the way of defining a set.

Barber's Paradox

Does the barber shave himself?

Liar Paradox

A liar says "I am lying".

Paradox

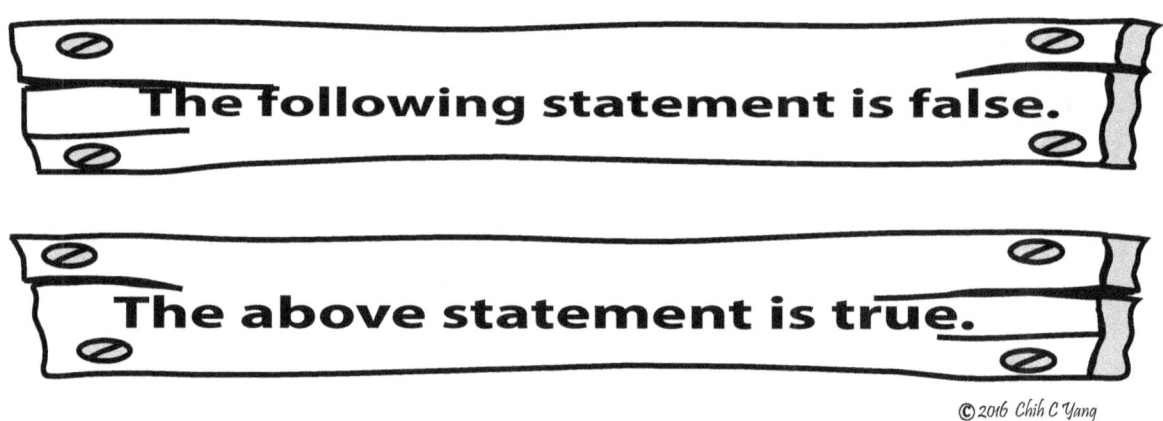

How to Avoid the Paradox?

The study of sets must begin with axioms that specify how sets can be constructed. For example, in the barber paradox, we can say that such a barber does not exist. Such paradoxes caused mathematicians to see the need of Axiomatic Formulation of sets. The Zermelo-Fraenkel set theory became the accepted axiomatic formulation of Cantor's idea, and remains so to this day. It is adequate for the needs of modern mathematics.

References – Set Theory

[1] Bond, Robert J. and William J. Keane, An Introduction to Abstract Mathematics, Brooks/Cole Publishing Co. CA. 1999

[2] Boyer, Carl B.; Merzbach, Uta C., A History of Mathematics, 2nd Edition. John Wiley & Sons. NY. 1991

[3] Breuer, Joseph, Introduction to the Theory of Sets, translated by Howard F. Fehr, Dover Publications, Inc., NY. 2006

[4] Bunch, Bryan, Mathematical Fallacies and Paradoxes, Dover Publications, Inc., NY 1982

[5] Devlin, Keith J., Sets, Functions, and Logic: An introduction to abstract mathematics, 3rd Edition. Chapman & Hall/CRC, FL. 2004

[6] Ferreiros, Jose, Labyrinth of Thought, A History of Set Theory and Its Role in Modern Mathematics, 2nd Edition. Birkhauser Verlag AG, Germany. 2007

[7] Kramer, Edna E., The Nature and Growth of Modern Mathematics, Princeton University Press, NJ. 1981

[8] Manor, Eli, To Infinity and Beyond: A Cultural History of the Infinite, Princeton University Press, NJ. 1987

[9] Niven, Ivan, Numbers: Rational and Irrational, Random House, Inc., NY. 1961

[10] Stewart, Ian, Concepts of Modern Mathematics, Dover Publications, Inc. NY. 1995

[11] Taylor, Angus E. and W. Robert Mann, Advanced Calculus, 2nd Edition. John Wiley & Sons. NY. 1972

5
Mathematical Induction

Carl Friedrich Gauss (1777 -1855)

When Gauss was ten years old he attended an arithmetic class. The teacher had given the class a difficult summation problem to keep them busy.

Gauss invented a shortcut formula and immediately wrote down the correct answer.

How did Gauss solve the problem?

Let the statement **S** denote the sum of all the numbers from 1 to 100, and arranged as:

$$1 + 2 + 3 + 4 + 5 + 6 + \ldots + 100 = \mathbf{S}$$
$$+)\ 100 + 99 + 98 + 97 + 96 + 95 + \ldots + 1 = \mathbf{S}$$
$$\overline{101 + 101 + 101 + 101 + 101 + 101 + \ldots + 101 = 2\mathbf{S}}$$

Thus, the answer is:
S = (100 + 1) × 100 / 2
   = 5050

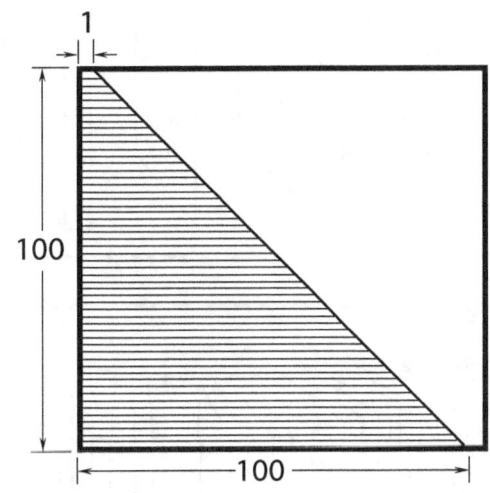

(Note: The scale is not proportional.)

New question

What happens when the term is changed to any number n?

$$\mathbf{S_2} = 1 + 2 + 3 + 4 + 5 + 6 + \ldots + n = ?$$

With the same approach, we got:

$$\mathbf{S_2} = 1 + 2 + 3 + 4 + 5 + 6 + \ldots + n$$
$$= (n + 1)\,n/2.$$

By this approach, the formula $\mathbf{S_2} = (n+1)n/2$ was discovered and proved.

However, this is not the only way of proof. There exists a different method called **Mathematical Induction**.

Mathematical Induction

The method of Mathematical Induction is similar to knocking down a string of dominoes. The imagery is that we have infinitely many dominoes which are lined up in a queue. If the first domino falls, then all of them will fall.

The Principle of Mathematical Induction

Let $P(n)$ is a statement that depends on the positive integer n.

1. If the statement $P(1)$ is true for n=1, and
2. The statement $P(k)$ is assumed true for n=k.
3. Based on the assumption, if the statement $P(k+1)$ is proved to be true for n=k+1

then the statement $P(n)$ is true for all positive integers n.

Example 1

Is the statement $P(n)$ true for any positive integer n?

$$1 + 2 + 3 + \ldots + n = n(n+1)/2 \qquad (A)$$

The numbers generated by formula (A) are called **triangle numbers**.

n= 1   n = 2   n = 3   n = 4   n = 5

Proof

 Step 1. When n=1, $1 = 1(1+1)/2$
P(1) is true.

 Step 2. We assume that $P(k)$ is true, when n=k.

$$1 + 2 + 3 + \ldots + k = k(k+1)/2$$

Step 3. Next, let n=k+1

We need prove that

$$1 + 2 + 3 + \ldots + k + (k+1) = (k+1)[(k+1)+1]/2$$

Because it is assumed that
$$1 + 2 + 3 + \ldots + k = k(k+1)/2$$

then

$$
\begin{aligned}
1 + 2 + 3 + \ldots + k + (k+1) &= [1 + 2 + 3 + \ldots + k] + (k+1) \\
&= [k(k+1)/2] + (k+1) \\
&= [(k^2 + k)/2] + 2(k+1)/2 \\
&= (k^2 + k + 2k + 2)/2 \\
&= (k^2 + 3k + 2)/2 \\
&= (k+1)(k+2)/2 \\
&= (k+1)[(k+1) + 1]/2
\end{aligned}
$$

Thus, P(k+1) is true.
Therefore, P(n) is true for all positive integers n.

Example 2

Is the statement P(n) true for any positive integer n?

$$1^2 + 2^2 + 3^2 + \ldots + n^2 = n(n+1)(2n+1)/6 \qquad (B)$$

The numbers generated by formula (B) are called **square pyramid numbers**.

n = 1   n = 2   n = 3   n = 4

  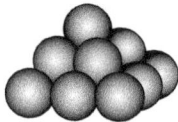 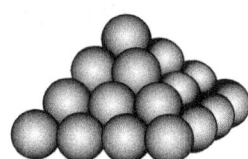

1      $1^2 + 2^2 = 5$ $1^2 + 2^2 + 3^2 = 14$ $1^2 + 2^2 + 3^2 + 4^2 = 30$

There are two ways to prove this. One is to build the formula logically from many simple algebraic identities, like Levi ben Gershon's (1288-1344) method, which is long-winded and complex. The other is through mathematical induction.

Proof

Mathematical Induction

 Step 1. When $n = 1$, $1^2 = 1(1+1)(2 \cdot 1 + 1)/6 = 1$
$P(1)$ is true.

 Step 2. We assume that $P(k)$ is true, when $n = k$.
$$1^2 + 2^2 + 3^2 + \ldots + k^2 = k(k+1)(2k+1)/6$$

 Step 3. Next, let $n = k+1$

We need prove that

$$\boxed{1^2 + 2^2 + 3^2 + \ldots + k^2 + (k+1)^2 = (k+1)[(k+1)+1][2(k+1)+1]/6}$$

Because it is assumed that
$$1^2 + 2^2 + 3^2 + \ldots + k^2 = k(k+1)(2k+1)/6$$

then

$$\boxed{1^2 + 2^2 + 3^2 + \ldots + k^2 + (k+1)^2}$$

$= k(k+1)(2k+1)/6 + (k+1)^2 \quad = (k+1)\,[k(2k+1)/6 + (k+1)]$
$= (k+1)(2k + 7k + 6)/6 \quad = (k+1)\,(k+2)\,(2k+3)/6$

$$= \boxed{(k+1)[(k+1)+1][2(k+1)+1]/6}$$

Thus, $P(k+1)$ is true.
Therefore, $P(n)$ is true for all positive integers n.

Mathematical induction is simple and often works smoothly. It is an important proof technique that can be used to prove the type of statements similar to the examples. However, it is used only to prove results obtained from some other ways, *rather than a tool for discovering formulae.*

Tower of Hanoi

The tower of Hanoi is a game invented by French mathematician E. Lucas in 1883. It consists of three pegs and several different sized disks.

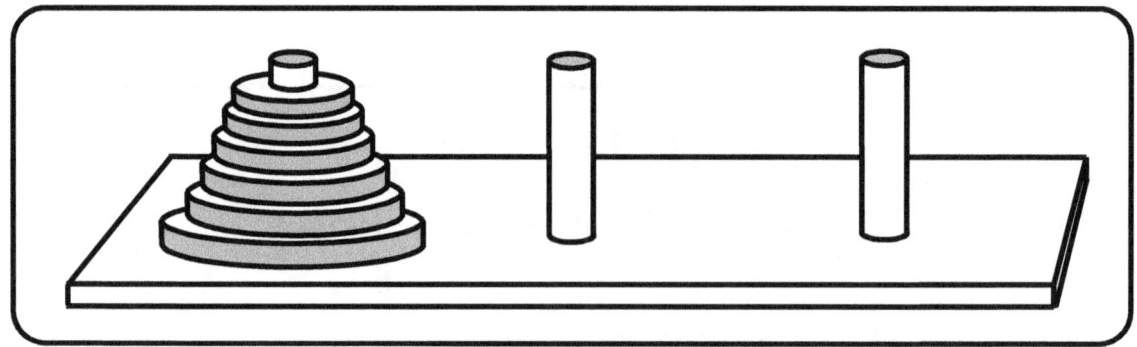

The game is to transfer all the disks to another peg.

Rule #1 Move only one disk at a time.
Rule #2 A disk may be moved to any peg.
Rule #3 Never place a disk on top of smaller disk.
Rule #4 The final order of the disks on the new peg
             must be the same as the original order.

The Tower of Hanoi with 3 Disks - It takes 7 steps.

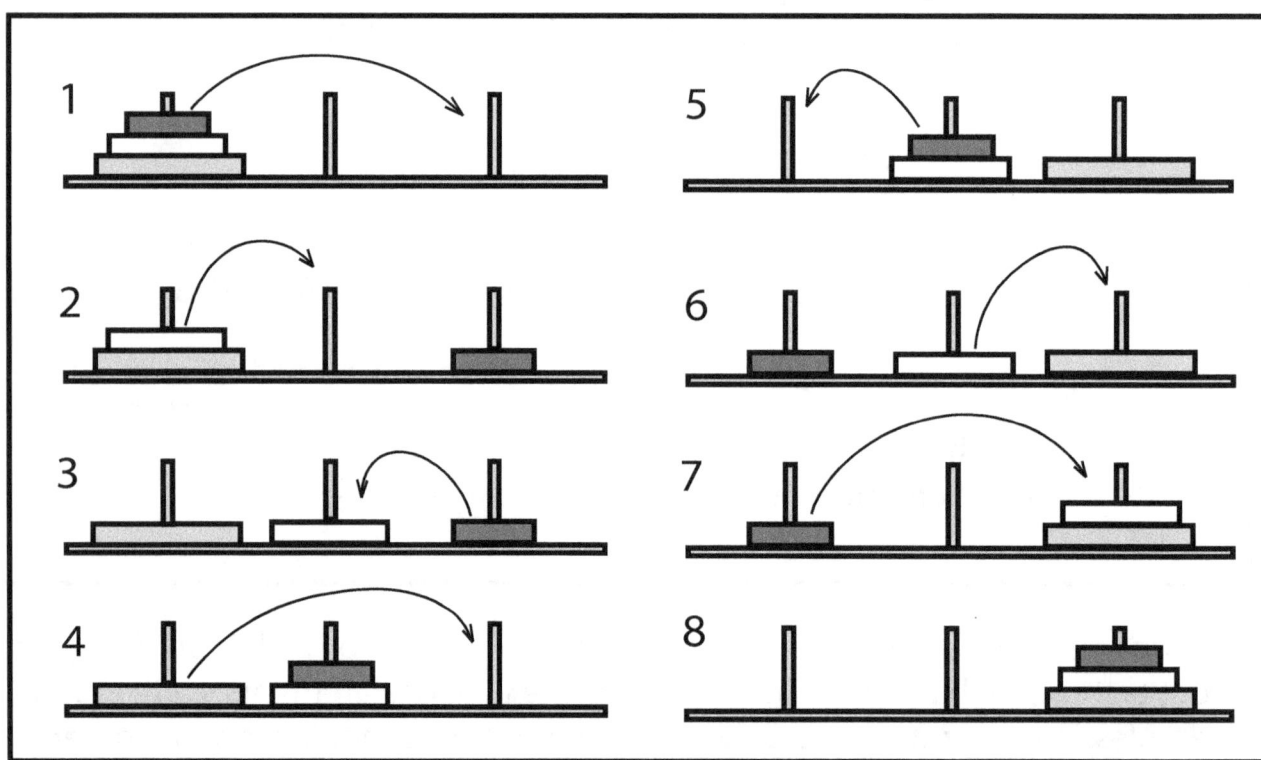

The Tower of Hanoi with 4 Disks - It takes 15 steps.

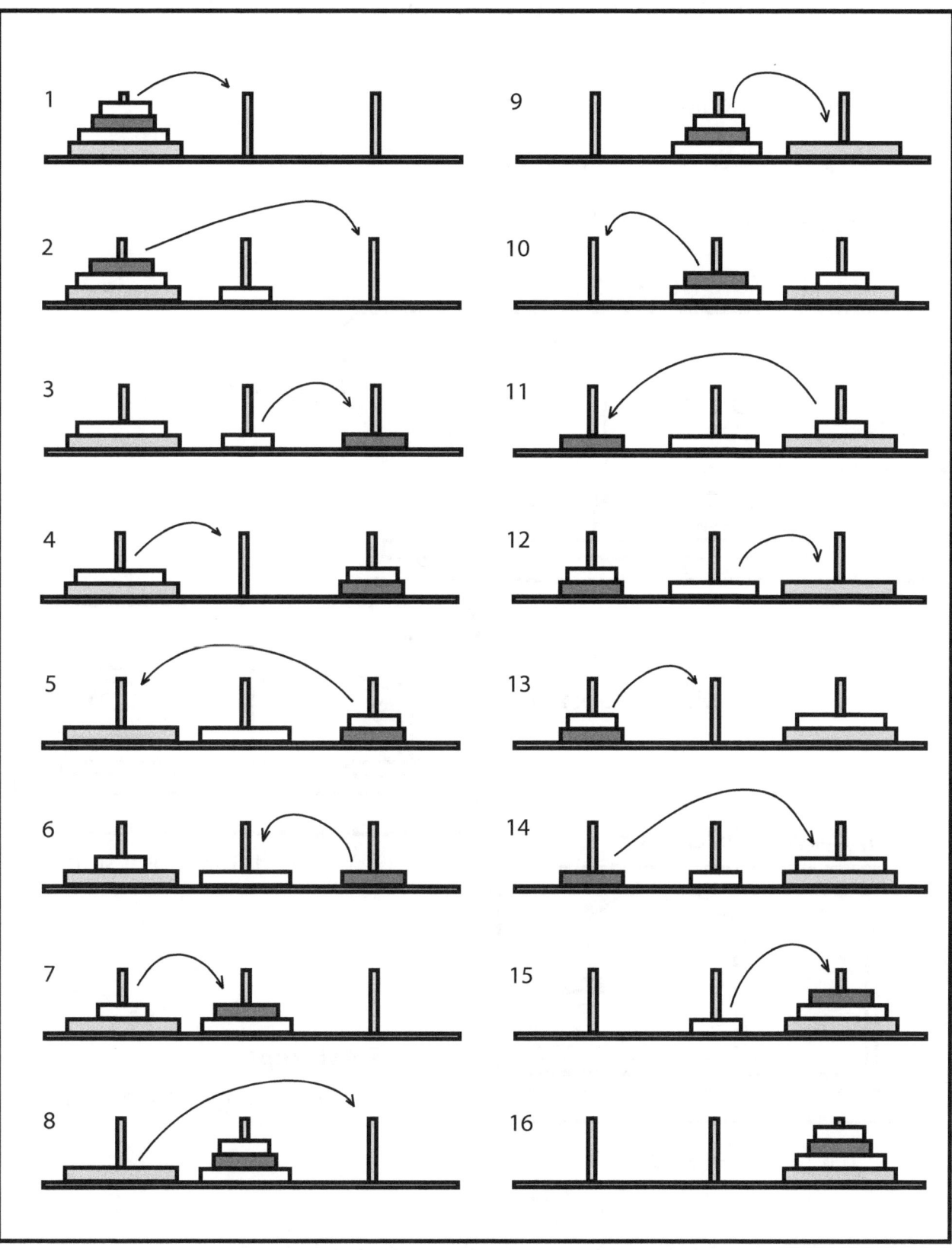

The Tower of Hanoi with *n* Disks

By mathematical induction, prove that the game can be completed in $2^n - 1$ steps.

Proof:

Step 1. When $n = 1$, it takes 1 step.

Step 2. When $n = k$, we assume it takes $2^k - 1$ steps.

Step 3. When $n = k + 1$

(1) Move k disks forward

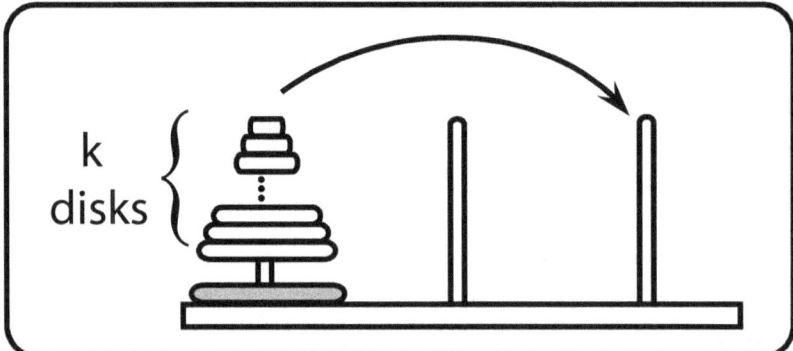

It takes $2^k - 1$ steps.

(2) Move the last one

It takes one step.

(3) Move k disks back

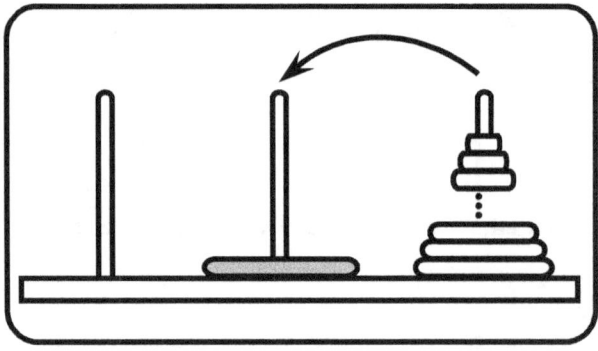

It takes $2^k - 1$ steps.

Total steps:
$(2^k - 1) + 1 + (2^k - 1) = 2^{k+1} - 1$

It is true that the game can be completed in $2^n - 1$ steps.

The Tromino Puzzle

The Tromino Puzzle provides a geometric example of mathematical induction.

> **Golomb's Tromino Theorem**
>
> *If any square is removed from a $2^n \times 2^n$ chessboard, the remaining board can be completely covered by L-shaped trominoes.*

Example

A square is removed from a $2^2 \times 2^2$ ($n = 2$) chessboard. The remaining 15 squares can be completely covered by L-shaped trominoes. The removed one could be any of the 16 squares.

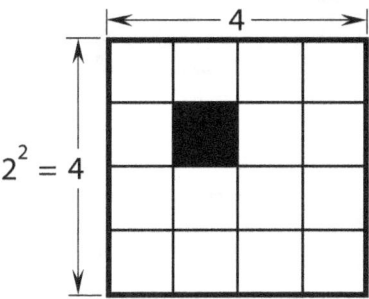

There are four orientations of L-shaped trominoes to cover the board.

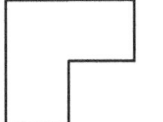 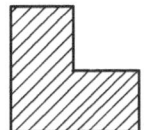 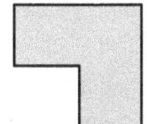 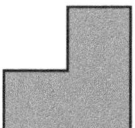

The following pictures show the covered boards with different locations of the removed square:

(1)   (2)   (3)   (4)

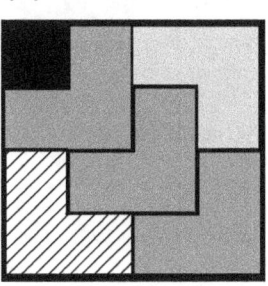 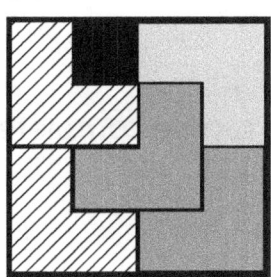 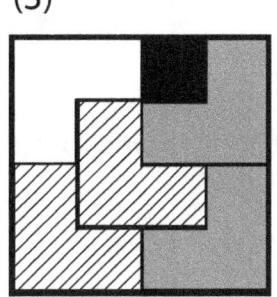 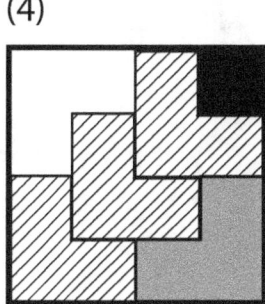

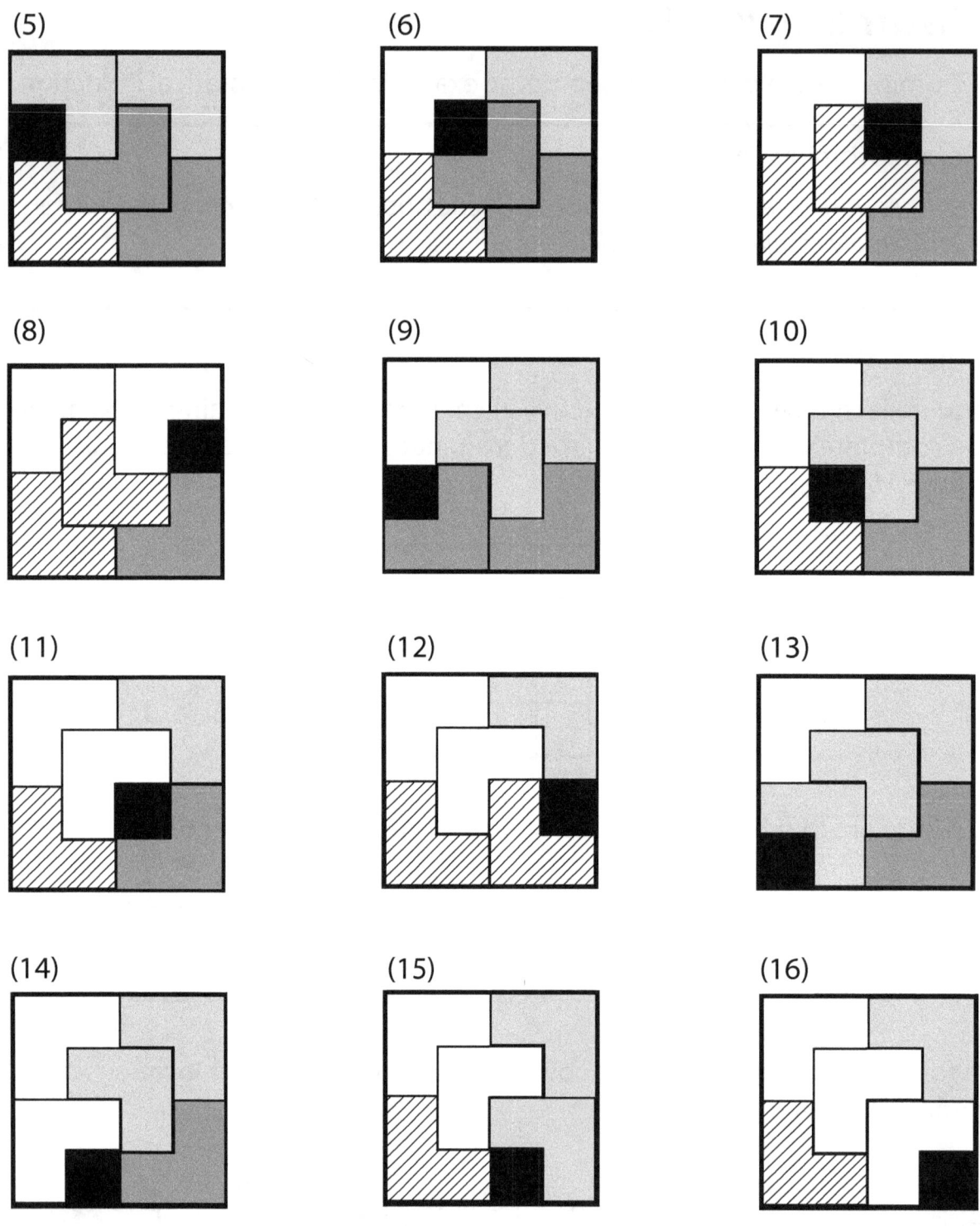

It is true that the remaining board can be completely covered.

The Proof of Golomb's Tromino Theorem

Step 1.

When $n=1$, the board has a size of 2x2.

One square is removed.

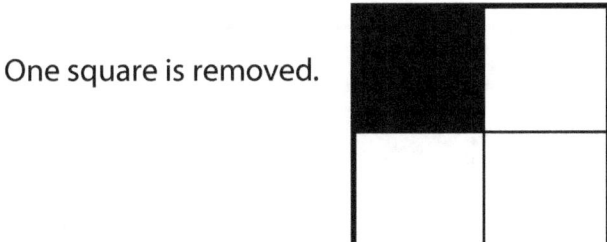

The covered board with four locations of the removed square:

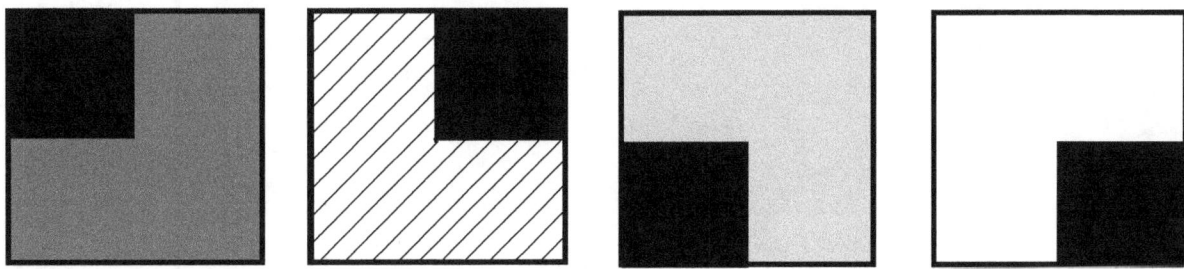

It is true that the remaining three squares can be covered by a L-shaped tromino.

Step 2.

When $n=k$, it is assumed true that a $2^k \times 2^k$ board with one square missing can be completely covered by L-shaped trominoes.

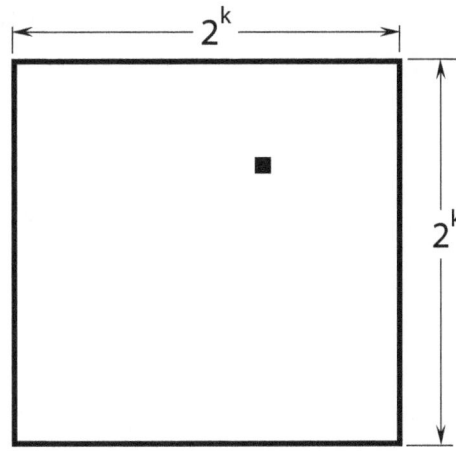

Step 3.

When n = k + 1, the chessboard's area is enlarged.

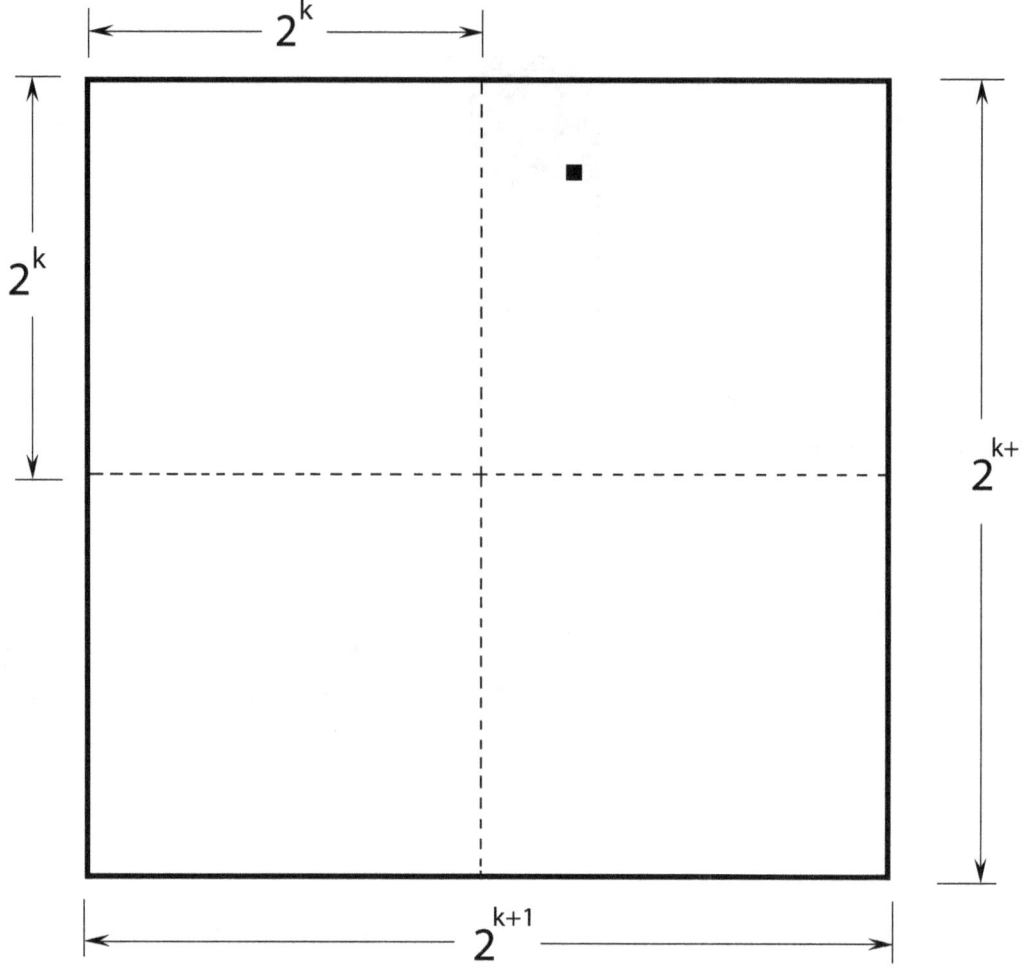

The $2^{k+1} \times 2^{k+1}$ chessboard has the size of four $2^k \times 2^k$ boards.

The $2^{k+1} \times 2^{k+1}$ board is made up of four $2^k \times 2^k$ quadrants. Quadrant *1* is a $2^k \times 2^k$ board with one square removed. It can be covered completely by L-shaped trominoes.

Next, a corner square is removed from each of the other three quadrants as shown below:

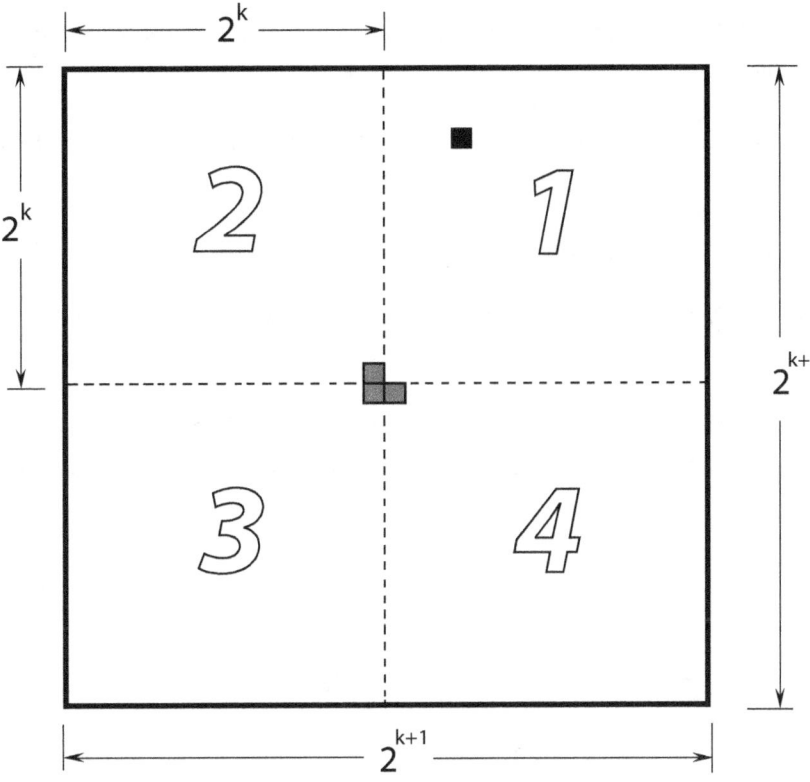

The 3 corner squares form a L-shaped tromino.

Then, quadrants *2*, *3*, and *4* each has a square missing. They can be completely covered by L-shaped trominoes. Thus, the entire board can be completely covered by L-shaped trominoes.

References – Mathematical Induction

[1] Bajnok, Bela; An Invitation to Abstract Mathematics, Springer, NY. 2013

[2] Bunch, Bryan; Mathematical Fallacies and Paradoxes, Dover Publications, Inc., NY. 1997

[3] Fletcher, Peter; Hoyle, Hughes; and Patty, C. Wayne; Foundations of Discrete Mathematics. PWS-Kent Publishing Company, Boston. 1991

[4] Hardy, Darel W. and Walker, Carol L.; Applied Algebra, Codes, Ciphers, and Discrete Algorithms. Pearson Hall, NJ. 2003

[5] Maddox, Randall B.; A Transition to Abstract Mathematics, Learning Mathematical Thinking and Writing, 2nd Edition. Elsevier Academic Press, UK. 2009

[6] Rosen, Kenneth H.; Discrete Mathematics and Its Applications, 4th Edition. WCB/McGraw-Hill, 1999

[7] Rotman, Joseph J.; A First Course in Abstract Algebra with Applications, 3rd Edition. Pearson Prentice Hall, NJ, 2006

[8] Simonson, Shai and Brian Hopkins; A Rabbi, Three Sums, and Three Problems, Resources for Teaching Discrete Mathematics, MAA Notes #74, Mathematical Association of America, Washington, DC, 2009

6
The Structure of Algebra

The Beginning of European Algebra

In the 12th century, al-Khwarizmi's works on the Hindu-Arabic numerals and the concept of algebra was brought to Europe.

The method of al-Khwarizmi's algebra, al-jabr, is "balancing".

$$2x^2 + x = 5 - 2x$$

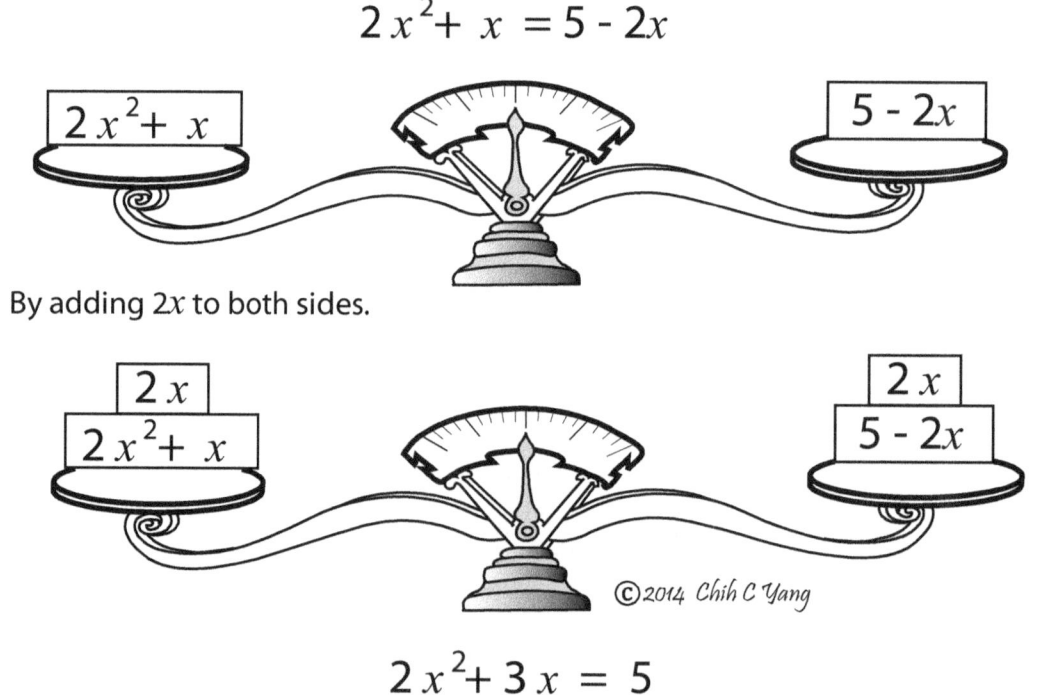

By adding $2x$ to both sides.

$$2x^2 + 3x = 5$$

He believed that a complicated math problem could be solved by breaking it down to several small steps.

The Rise of European Algebra

During the mid-16th century in Italy, math problem-solving competitions were a pastime and a way to display talent. Heavy bets were made. For a good contestant, these competitions were a source of income.

One of the most famous contestants was Gerolamo Cardano. He was a mathematician, physician, astrologer, gambler, and scoundrel.

Girolamo Cardano (1501 -1576)

Cardano wrote a book about the chance of gambling. It was an early treatment of probability, which included cheating techniques of gambling.

The first show of imaginary number $\sqrt{-1}$

In a math problem:

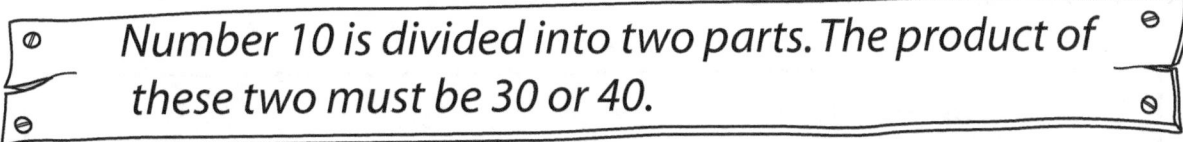
Number 10 is divided into two parts. The product of these two must be 30 or 40.

> Cardano came out with the answers
> $(5+\sqrt{-5})(5-\sqrt{-5}) = 30$
> or $(5+\sqrt{-15})(5-\sqrt{-15}) = 40$.
> But he believed the numbers $\sqrt{-5}$ and $\sqrt{-15}$ were non-rigorous.

By the end of the 16th century, algebra had emerged as a major branch of mathematics. Methods were found to solve different equations, in particular, all quartic and cubic equations. The challenge ahead was to find out the solution of equations with a degree of 5 or higher.

During the next 200 years, no one succeeded in finding the formula for the roots of equations of degree 5 or higher.

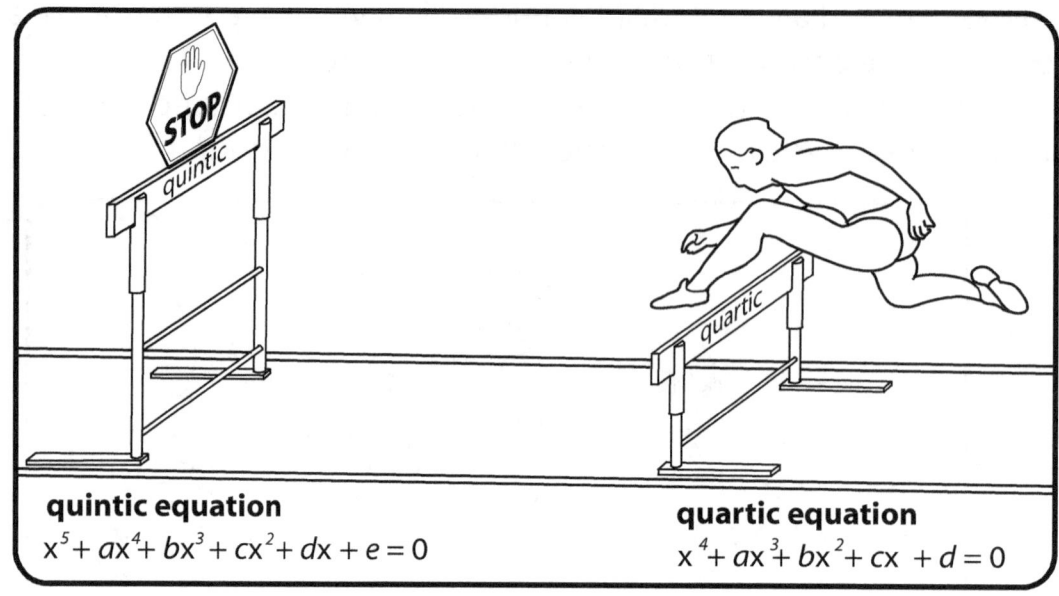

quintic equation
$x^5 + ax^4 + bx^3 + cx^2 + dx + e = 0$

quartic equation
$x^4 + ax^3 + bx^2 + cx + d = 0$

©2017 Chih C Yang

The Birth of Modern Algebra

In the 19th century, Niels Henrik Abel (1802 - 1829) and Evariste Galois (1811-1832) studied algebra's structure and concluded: There is no formula for the roots of equations with a degree of 5 or higher.

Niels Henrik Abel (1802 - 1829)

Abel's Lemniscate

This conclusion signaled the beginning of modern algebra.

The Anatomy of Algebra

In the era of modern algebra, algebra is no longer conceived as the science of equation solving. Its new direction is the studying of algebra itself, algebraic structure.

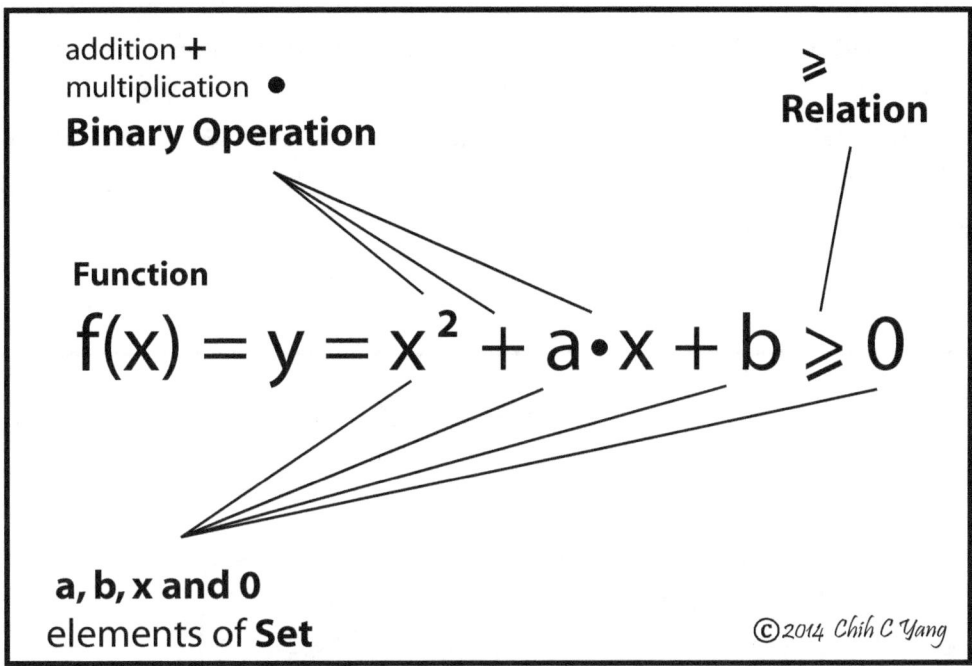

Algebra's basic structure consists of sets, operations and relations.

- ■ A **Set** is a collection of objects, not necessary limited to numbers.

- ■ A **Binary Operation** is a process defined on a set, which combines two elements to produce a third element.

- ■ **Relations** relate the result of an operation or **Function** to an object.

Evariste Galois (1811 - 1832)

Galois is famous for his contributions to algebra now known as group theory. His theory allow us to show that the general quintic equation cannot be solved by radicals. He died at the age of 20 in 1832 in a duel.

Function

Before the 19th century, mathematicians followed the conventional way of using a formula such as $y = x^2 - 4x + 3$ to calculate a number y from a given number x. In the era of modern mathematics, Dirichlet (1805 - 1859) replaced the formula with the concept of the **function**. A function, working like an "input-output" device, is a rule that produces new numbers from given numbers. The rule does not have to be an algebraic formula. And the inputs/outputs are not necessary limited to numbers.

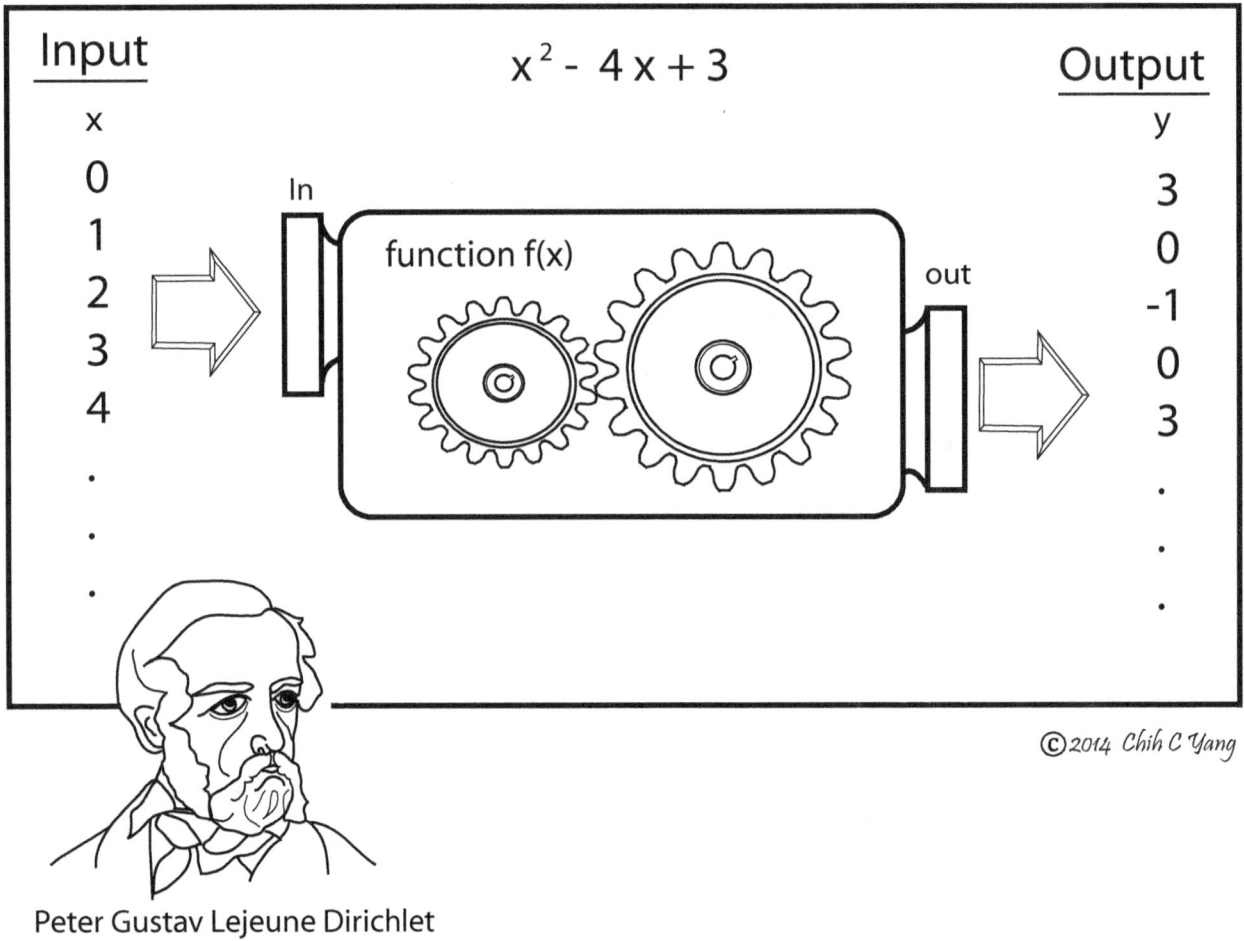

Peter Gustav Lejeune Dirichlet
(1805 - 1859)

In modern algebra the terms **Mapping** or **Transformation** are also used in place of **function**.

Function

A function ϕ is a mapping from domain to codomain. A domain is the set of objects where the function is defined. A function will accept inputs only from its domain. The codomain is the target set into which all of the outputs are constrained to fall.

Mathematical Groundhog Day

Groundhog day is a festival celebrated on February 2 in the United States and Canada. According to folklore, when a groundhog emerges from its burrow on this day and sees its shadow, which means six more weeks of winter.

Function

A figure after a transformation (function ϕ) is called its image.
Each element has only one image.

Not a Function

A function can not have two or more images for the same element of its domain.

When the groundhog couldn't see its shadow, which predicts an early spring.

Not a Function

To be a function, every element in the domain should have an image.

Function

The Lone Ranger ϕ versus a gang of desperados

© 2017 Chih C. Yang

Domain

The Lone Ranger's Bullets

The ranger's math revolver

© 2017 Chih C. Yang

Let **B** be the set of ranger's bullets
B = {1, 2, 3, 4, 5}

Codomain

The Gangsters

Let **G** be the set of the gang members
G = {G1, G2, G3, G4}

Function $\phi : B \rightarrow G$

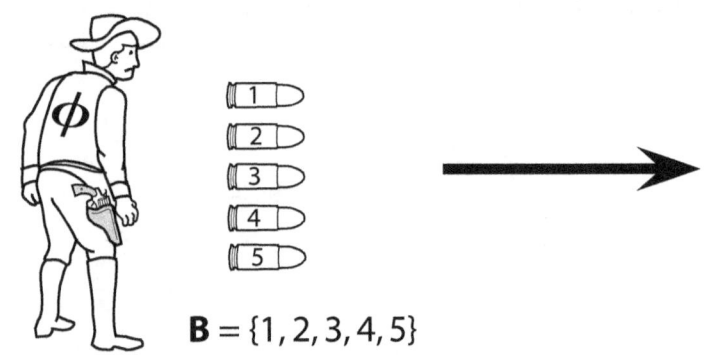

$B = \{1, 2, 3, 4, 5\}$

$G = \{G1, G2, G3, G4\}$

ϕ is a function from domain **B** to codomain **G**.

Example 1 Function

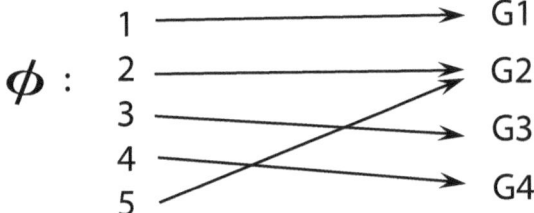

Onto
All gangsters got hit.

Example 2 Function

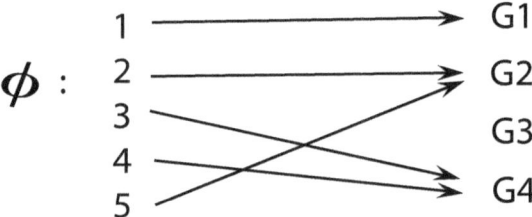

Not onto
Gangster G3 hid away.
Not all gangsters got hit.

Example 3 Not a function

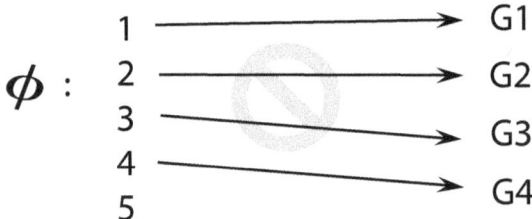

One bullet was left unused,

Example 4 Not a function

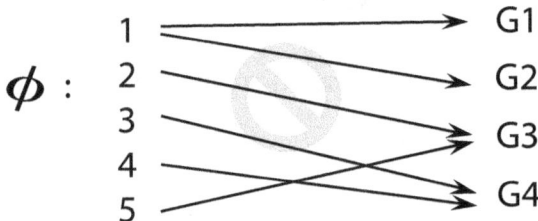

One bullet hit two gangsters.

Binary Operation

Operations like addition, subtraction, and multiplication are called binary operations. A binary operation is a process or rule used on a set that combines two elements from the set and produces a unique third element of the same set.

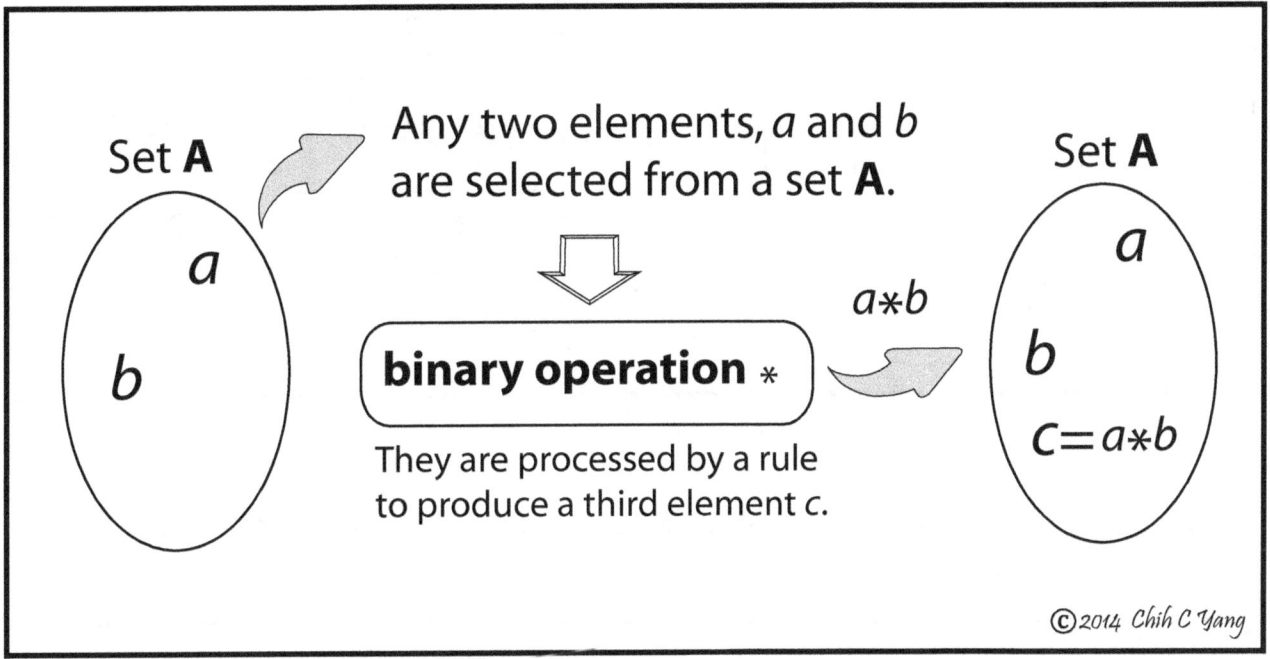

There are two parts of a binary operation:
(1) The set **A** must be **closed** under the operation.
(2) The operation must be **well-defined**.

■ **Closed**: After the operation, the produced third element c must still be in the same set **A** as a and b.

■ **Well-Defined**: The assigned element c must be **unique**. There can be one and only one value.

The Closure of a Binary Operation

For each a and $b \in \mathbf{A}$, $a * b \in \mathbf{A}$.

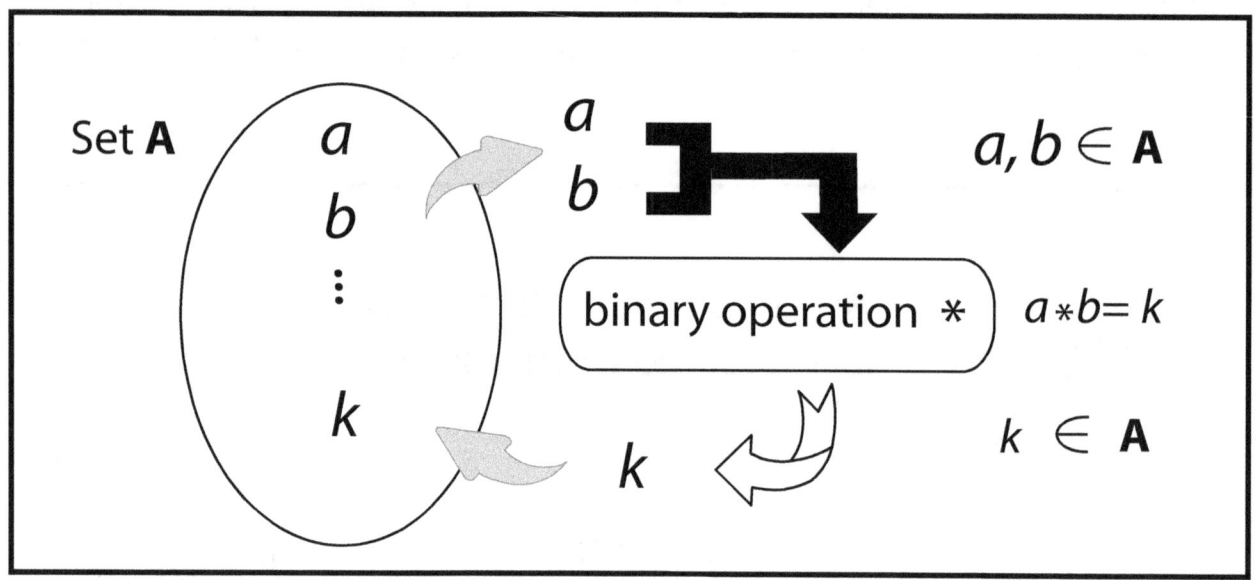

Example 1   The set of odd integers, **B**, is not closed under the operation of addition, +.

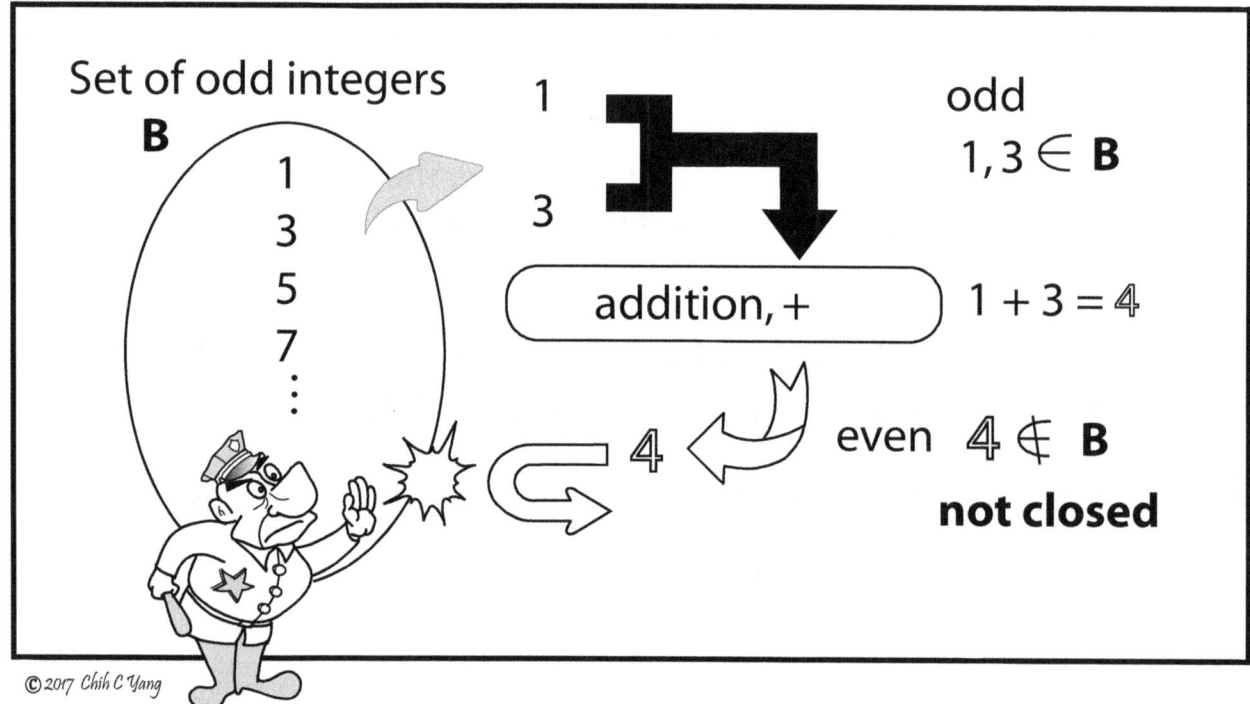

Example 2 The set of odd integers, **B**, is closed under the operation of multiplication.

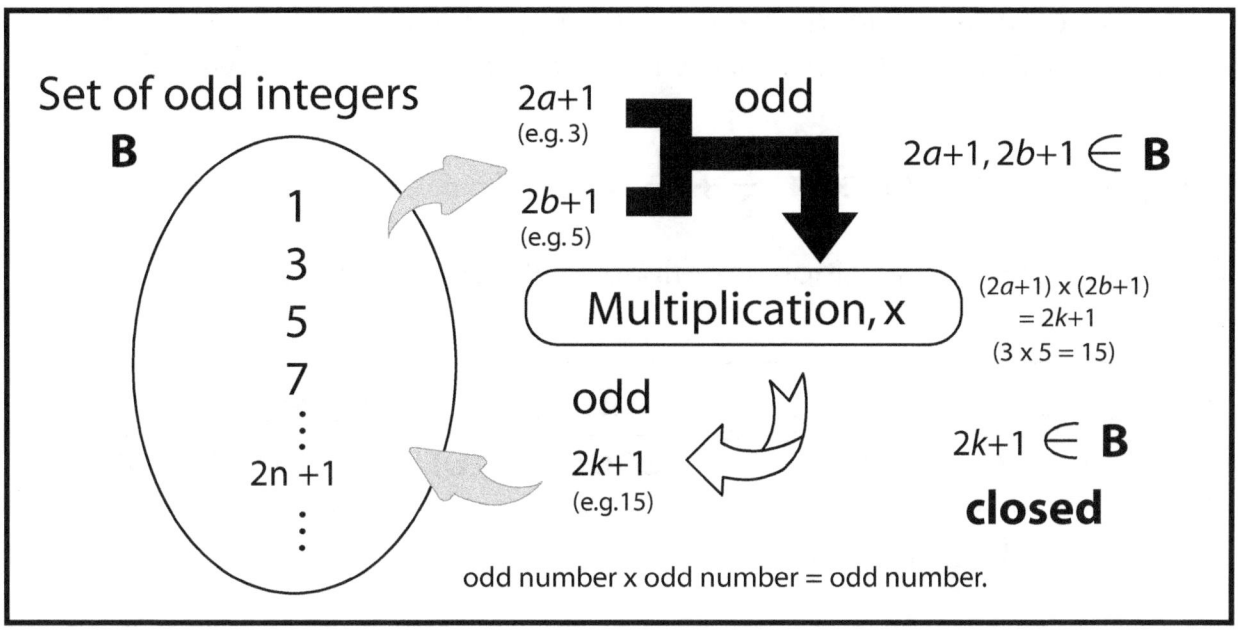

Like many sports, the operation must be confined to a particular area or domain.

closure law

penalty for going off the mat

A Well-defined Binary Operation

A binary operation defined on a set **A** is a mapping from set of ordered pairs of elements of **A** into **A**. It must follow the rule of mapping and have only one image.

Example 1   Why does $\frac{1}{2} + \frac{1}{3} \neq \frac{2}{5}$?

If a binary operation "addition" is defined on the set of rational numbers as shown:

$$\frac{a}{b} \oplus \frac{c}{d} = \frac{a+c}{b+d} = \frac{\text{numerator} + \text{numerator}}{\text{denominator} + \text{denominator}}$$

We get $\quad \frac{1}{2} \oplus \frac{1}{3} = \frac{2}{5}$

But $\quad \because \frac{1}{2} = \frac{2}{4} \quad$ and $\quad \frac{1}{2} \oplus \frac{1}{3} = \frac{2}{4} \oplus \frac{1}{3} = \frac{3}{7}$

So $\quad \frac{2}{5} = \frac{3}{7} \qquad$ ...It has more than one image.
$\qquad\qquad\qquad\qquad$ The operation is **not well-defined**.

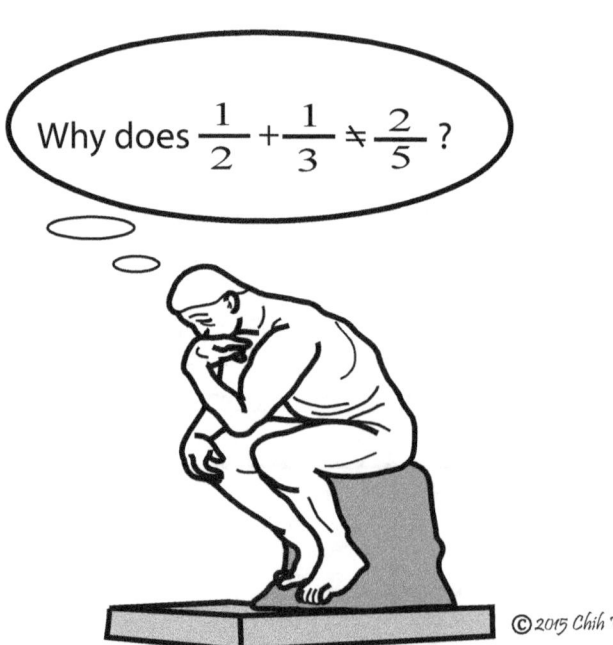

Example 2

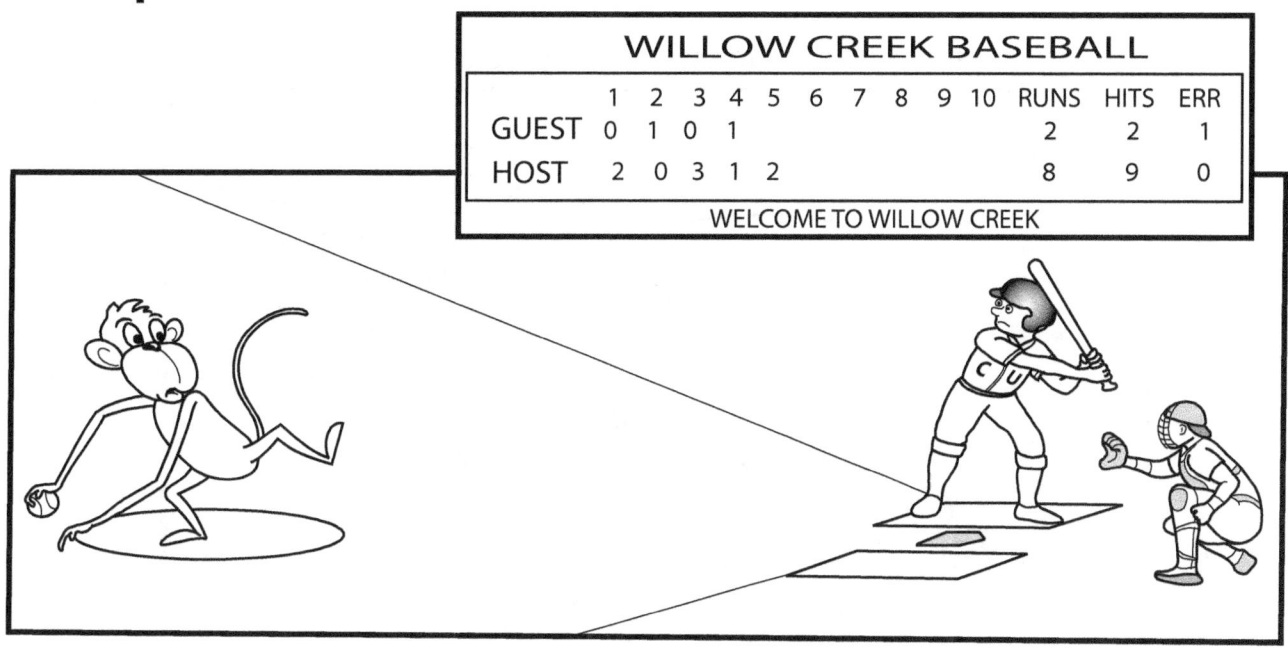

©2017 Chih C Yang

The Willow Creek baseball team has a game record for the past three seasons of 7 wins out of 13, 6 wins out of 12 and 8 wins out of 14 games.

If we follow the binary operation of example 1:

$$\frac{a}{b} \oplus \frac{c}{d} = \frac{a+c}{b+d} = \frac{\text{numerator} + \text{numerator}}{\text{denominator} + \text{denominator}}$$

Add them up: $\quad \dfrac{7}{13} \oplus \dfrac{6}{12} \oplus \dfrac{8}{14} = \dfrac{21}{39}$

There is a total of 21 wins out of 39 games.

Why does this work?
(Appendix for details)

The Order of Operations

Unary operations vs. Binary operations

The unary operation is an operation with only one operand, such as negation, factorial, or square root.

Example

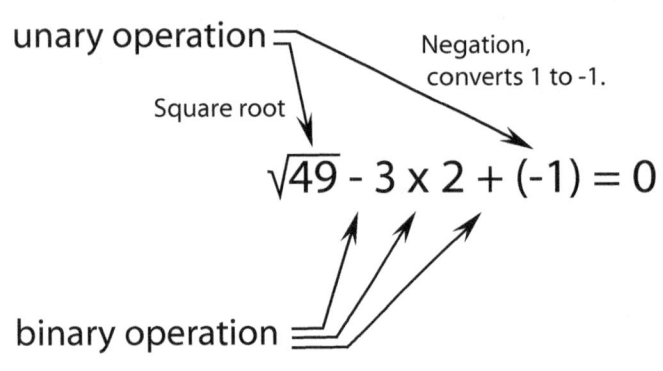

$\sqrt{49} - 3 \times 2 + (-1) = 0$

Operator	Operation	Precedence
( )	Parentheses	1st
!, √ , negation	Unary	2nd
×, ÷	Binary	3rd
+, −	Binary	4th
=, <, >	Assignment (Relation)	5th

The unary operation is granted a higher precedence than binary operation. In binary operations, division and multiplication take precedence over both addition and subtraction.

Relation

Relations connect elements of the same set in various ways.
For examples,
- $1 < 2$
- $a = \sqrt{b}$
- Similar triangles ABC ~ DEF

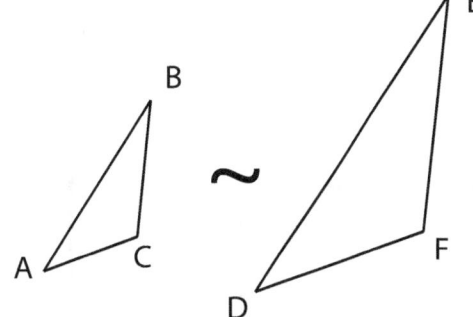

A **relation** on a set **A** is a set **R** of ordered pairs of elements of **A**.

Example 1 Relation on set **A**: matching different objects with similar shapes.

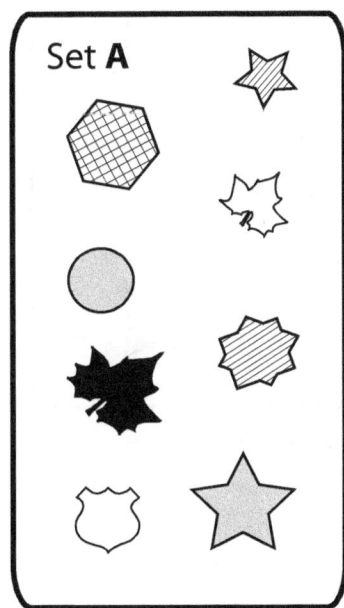

1. Pick any element from set **A**.
2. Match it with elements of similar shape in the same set **A**.

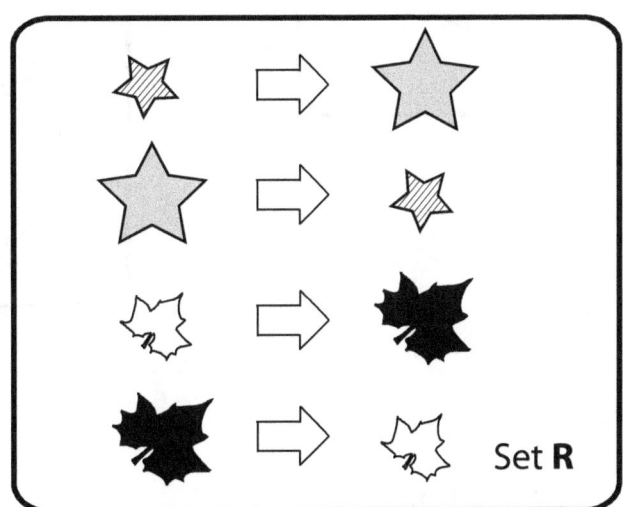

The order is one-way.

Relation **R**, a set of Ordered Pairs

Example 2 Relation on set **A**: matching different objects with the same shade.

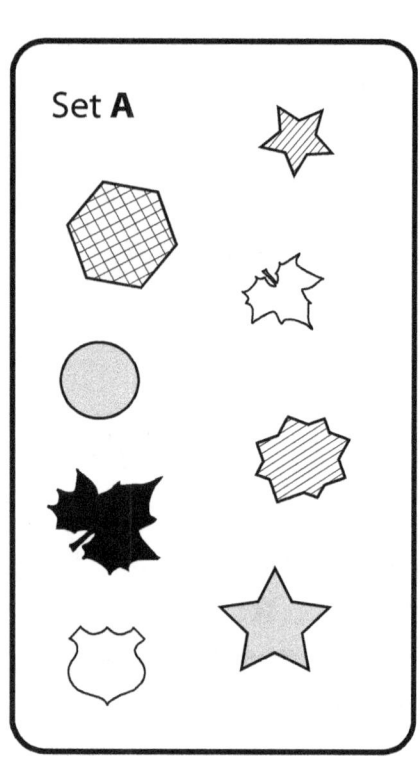

Ordered Pairs

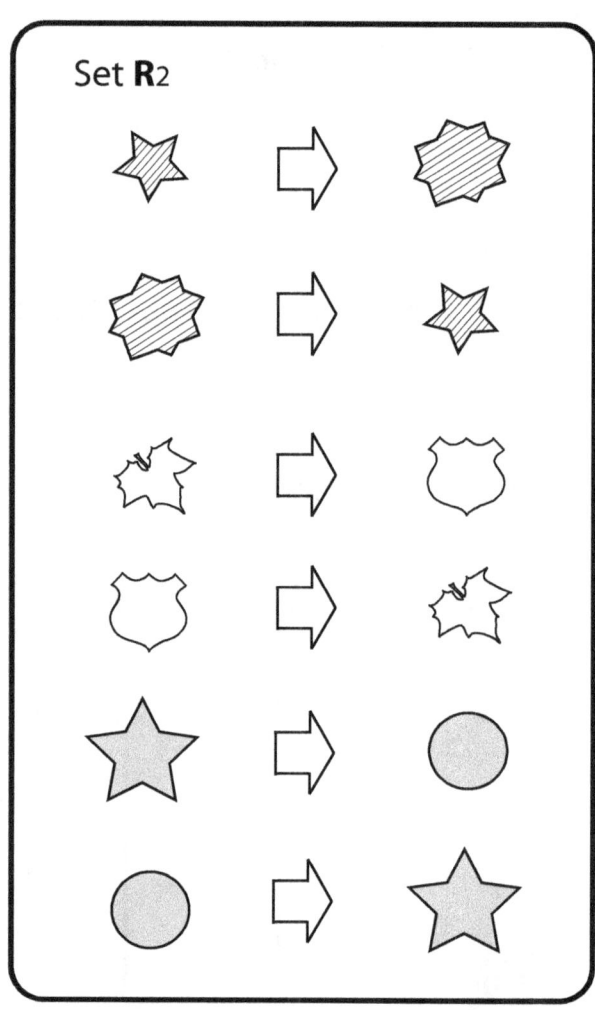

The order is one-way.

The relation

$\mathbf{R}_2 = \{(\text{☆}, \text{✸}), (\text{✸}, \text{☆}), (\text{🍁}, \text{🛡}), (\text{🛡}, \text{🍁}), (\text{★}, \text{●}), (\text{●}, \text{★})\}$

Example 3 A relation R_3 on set **A**: double an element x, $2x = y$.

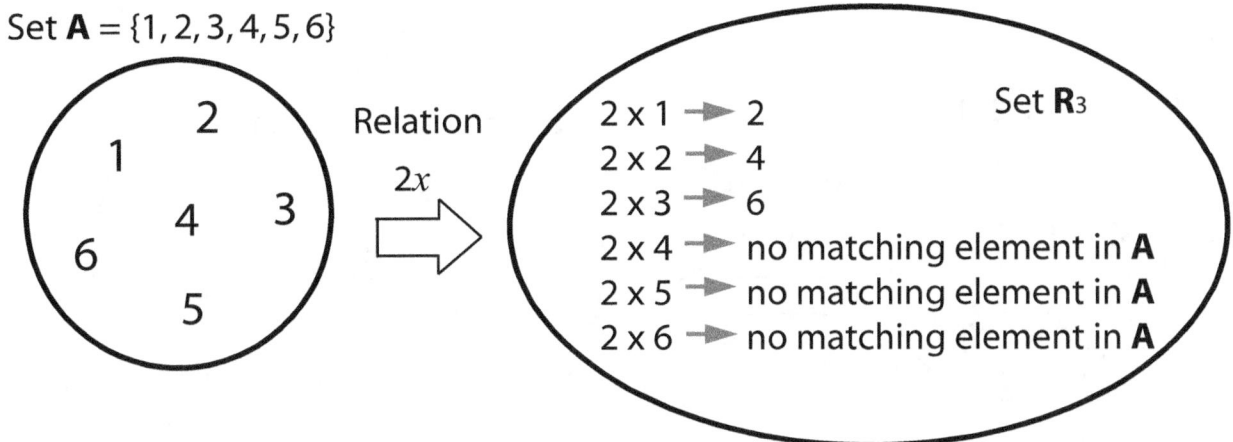

Set **A** = {1, 2, 3, 4, 5, 6}

Relation R_3 = {(1, 2), (2, 4), (3, 6)}
They can be written as $1R2$, $2R4$, and $3R6$.
$xRy \Longleftrightarrow 2x = y$

Set **AxA** = {(1, 1), (1, 2), (1, 3) ... (6, 4), (6, 5), (6, 6)}
There are 6 x 6 = 36 possible arrangements for set **AxA**.
R_3 is a subset of **AxA**, $R_3 \subseteq$ **AxA**

Note: Set **AxA** is a Cartesian product of **A**. That is, for set **A**, the Cartesian product **AxA** is the set of all ordered pairs (a, b) where $a, b \in$ **A**.

Directed Graphs

Directed graphs provide a good tool to describe relations. For example, the relation $1R2$ is illustrated as:

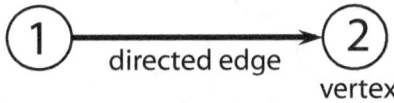

The relation R_3 = {(1, 2), (2, 4), (3, 6)}, can be illustrated as:

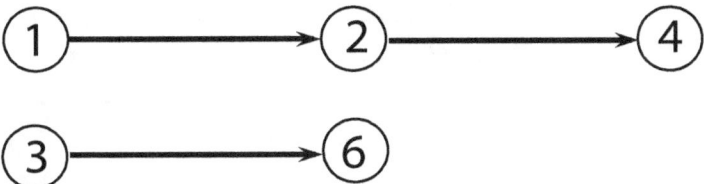

Example 4 A relation R_4 on set **B**: $x < y$

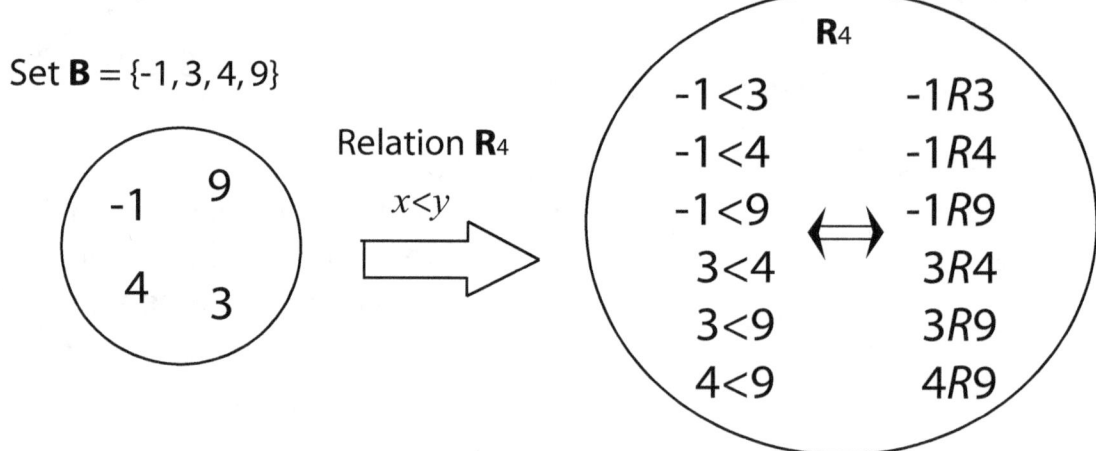

Set **B** = {-1, 3, 4, 9}

Relation R_4: $x<y$

R_4
-1<3   -1R3
-1<4   -1R4
-1<9   -1R9
3<4   3R4
3<9   3R9
4<9   4R9

Relation R_4 = {(-1, 3), (-1, 4), (-1, 9), (3, 4), (3, 9), (4, 9)}

Written as: -1R3, -1R4, -1R9, 3R4, 3R9, and 4R9

$xRy \iff x < y$

Relation R_4 is not *symmetric* since $3 < 4$, $(3, 4) \in R_4$, but $4 \not< 3$, $(4, 3) \notin R_4$.
Relation R_4 is not *reflexive* since $3 \not< 3$, $(3, 3) \notin R_4$.

Directed Graphs

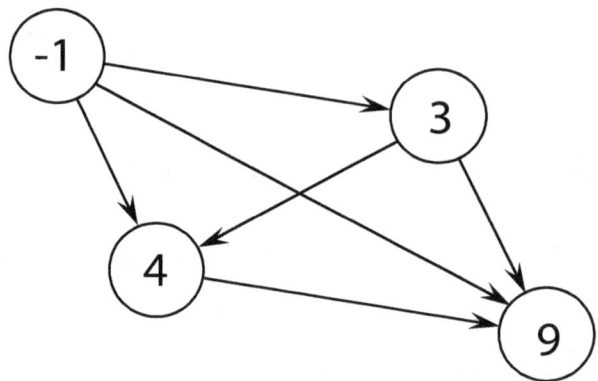

Example 5 A relation R_5 on set **C**: $|x| = |y|$

Set **C** = {-2, -1, 2, 5}

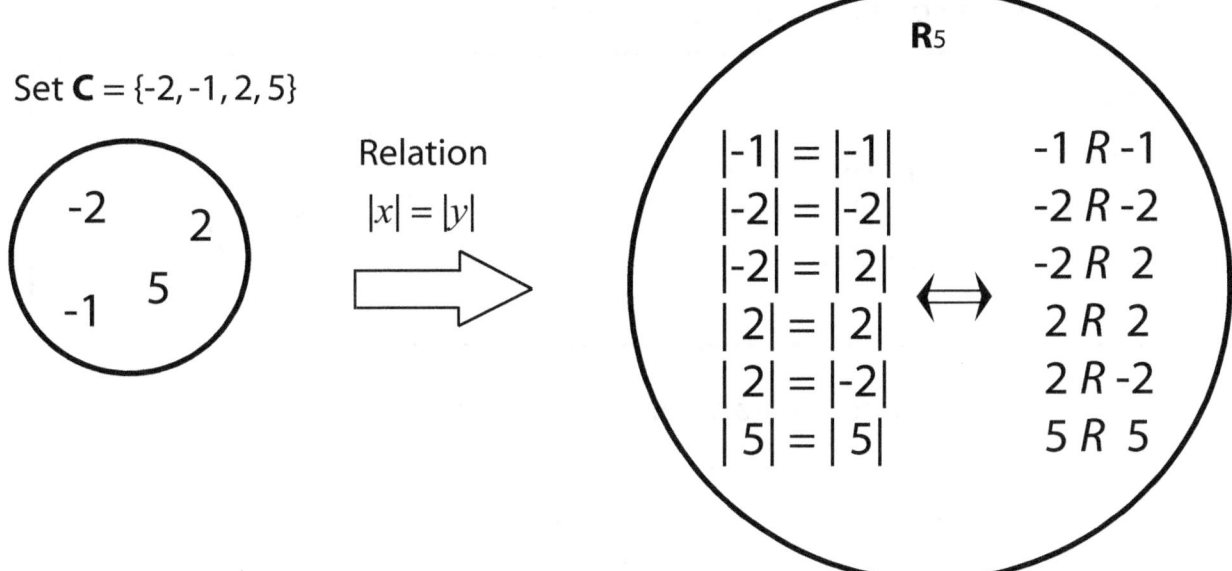

Relation R_5 = { (-1, -1), (-2, -2), (-2, 2), (2, 2), (2, -2), (5, 5)}

Written as: -1 R -1, -2 R -2, -2 R 2, 2R2, 2R -2, 5R5

$xRy \iff |x| = |y|$

Directed Graphs

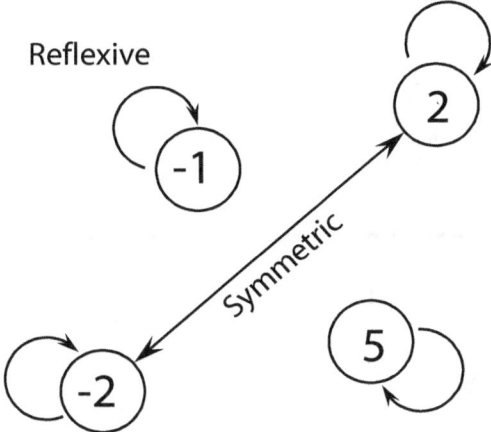

Relations (-1, -1), (-2, -2), (2, 2), and (5, 5) are called **Reflexive**.

$$|x| = |x|$$
$$xRx$$

Relations (2, -2) and (-2, 2) are called **Symmetric**.

If $|x| = |y|$ then $|y| = |x|$

if xRy then yRx

Properties of Relations

If a relation **R** is a on a set **A**, then

1. **R** is **reflexive** provided that xRx for all $x \in$ **A**.

> **Example**
>
> The relation of similar triangles, ~.
> Any triangle is similar to itself.
> ABC ~ ABC
>
>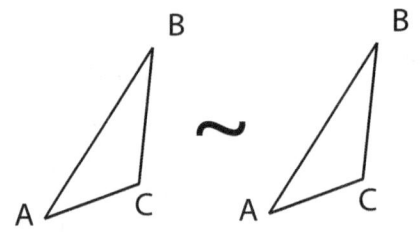

The directed graph of reflexive relation

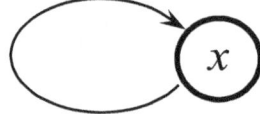

2. **R** is **symmetric** provided that xRy then yRx for all $x, y \in$ **A**.

> **Example 1** Similar triangles ABC ~ DEF $\Rightarrow$ DEF ~ ABC
>
>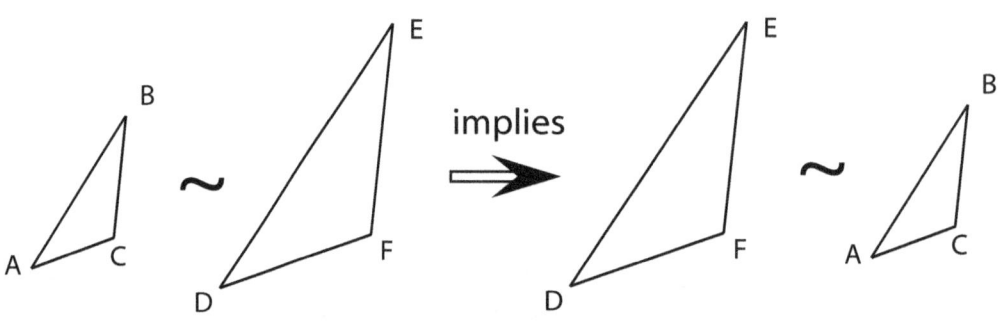
>
> **Example 2** Relation $|-2| = |2|$ implies $|2| = |-2|$
>
> $|x| = |y|$ implies $|y| = |x|$

The directed graph of symmetric relation

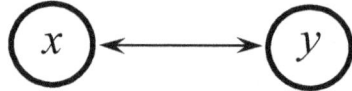

3. **R** is **transitive** provided that if xRy and yRz then xRz for arbitrary x, y, z in **A**.

Example 1

Similar triangles, $abc \sim DEF$ and $DEF \sim GHK$
then $abc \sim GHK$

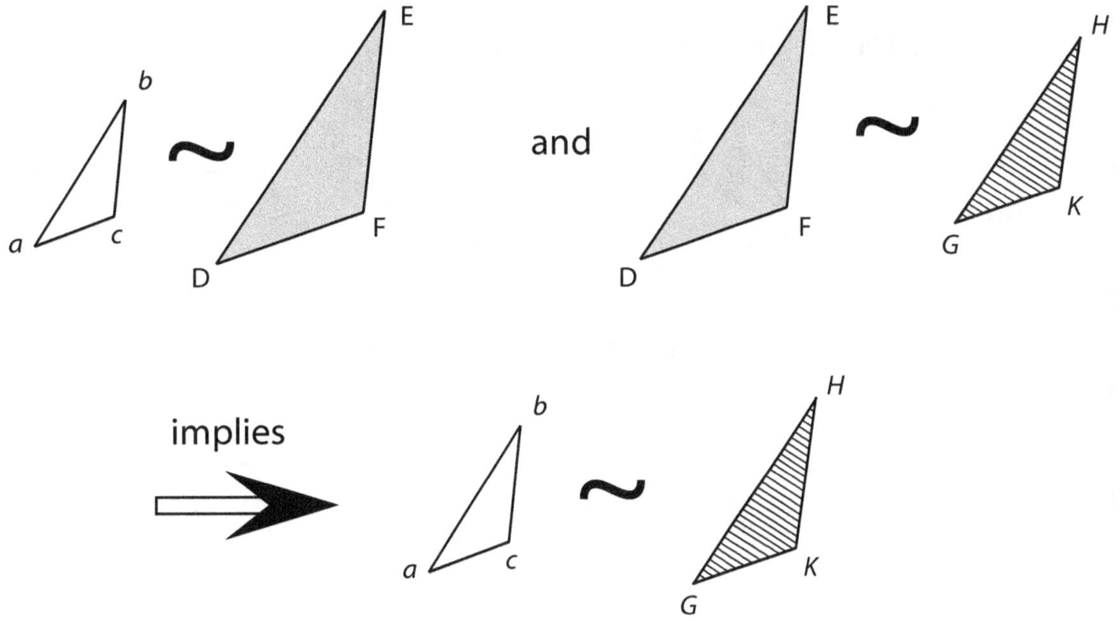

and

implies

Example 2

$3 < 5$ and $5 < 8$ then $3 < 8$

The directed graph of transitive relation

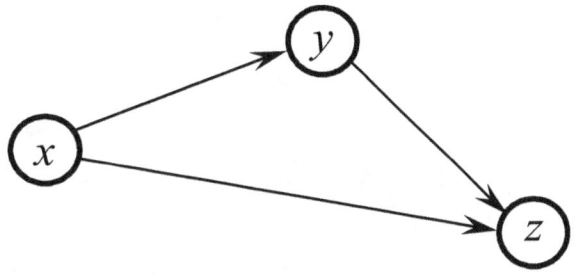

Equivalence Relation

A relation is an equivalence relation if it has all three properties: reflexive, symmetric and transitive.

Example 1

A relation of "similar shape" is defined on the set **A**.

Set **A** = {  }

The relation of similar shape is an equivalence relation.

Directed Graphs

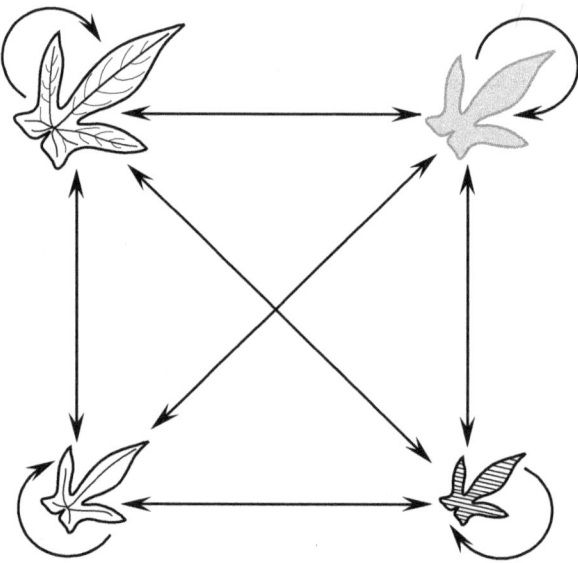

Equivalence Relation

Example 2

A relation **R** is defined on the set of integers **Z**.

$$aRb \iff |a| = |b|, \quad \text{any } a, b \in \mathbf{Z}$$

Reflexive $|a| = |a|$ aRa

Symmetric $|a| = |b|$ then $|b| = |a|$
  aRb then bRa

Transitive $|a| = |b|$ and $|b| = |c|$ then $|a| = |c|$, $c \in \mathbf{Z}$
  aRb and bRc then aRc

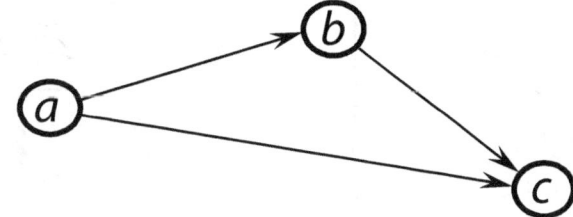

Equivalence relation

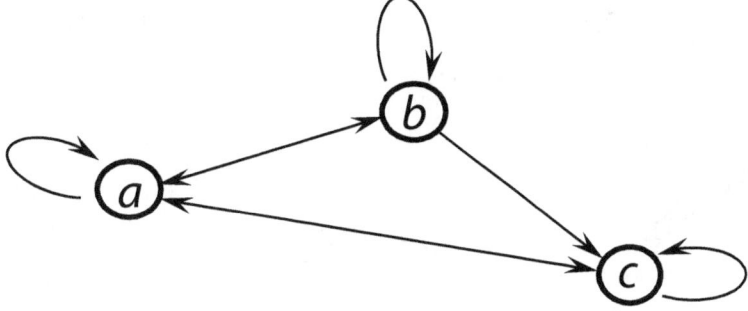

Example 3 Modular Arithmetic (Clock Arithmetic)

In modular arithmetic, operations are performed along a circle instead of a straight number line.

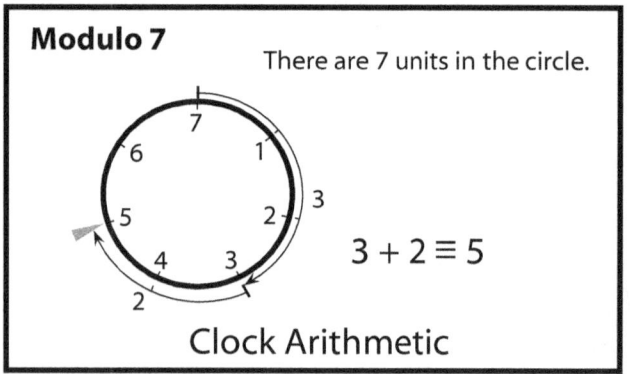
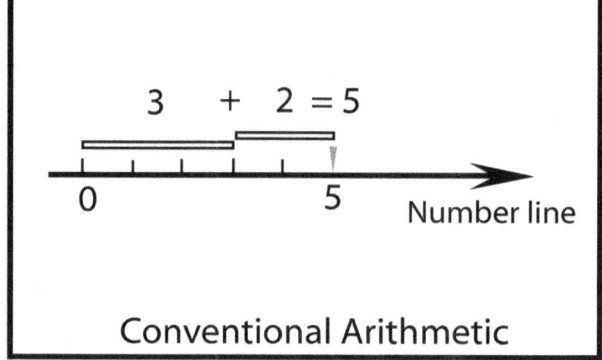

Example: The 12-hour Clock (Modulo 12)

11 hours after 2:00

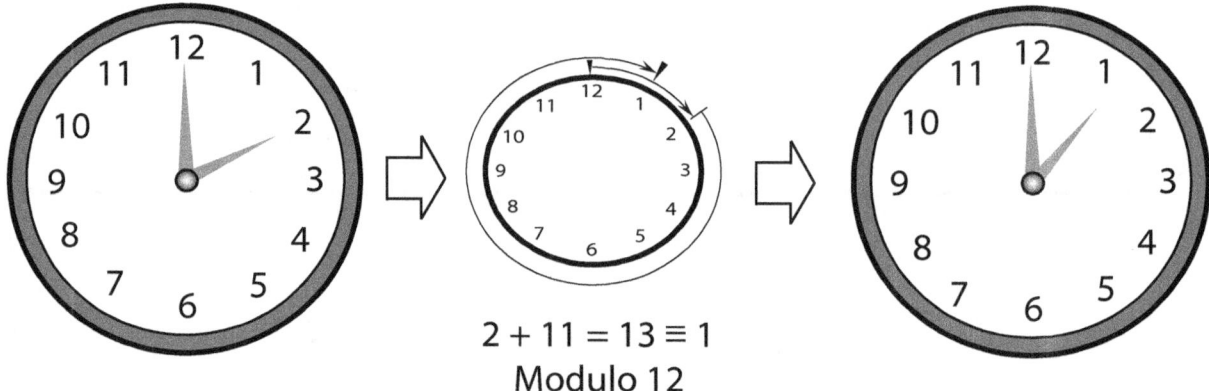

$2 + 11 = 13 \equiv 1$
Modulo 12

If it is 2 o'clock now, then 11 hours later it will be 1 o'clock. Every time we go past 12 on the clock we start counting the hour at 1 again. In this example, 1 and 13, at the same clock position, are **congruent** to each other, $1 \equiv 13$.

Let **A** be a set of the daily 24 hours. **A** = {1, 2, 3, ..., 22, 23, 24}

A relation **R** is defined on the set of **A**. For $a, b \in$ **A**,

$$a R b \iff a \equiv b \pmod{12}$$

1 R 13		1 ≡ 13
2 R 14		2 ≡ 14
3 R 15		3 ≡ 15
4 R 16		4 ≡ 16
5 R 17		5 ≡ 17
6 R 18	⟺	6 ≡ 18
7 R 19		7 ≡ 19
8 R 20		8 ≡ 20
9 R 21		9 ≡ 21
10 R 22		10 ≡ 22
11 R 23		11 ≡ 23
12 R 24		12 ≡ 24

Congruence modulo 12

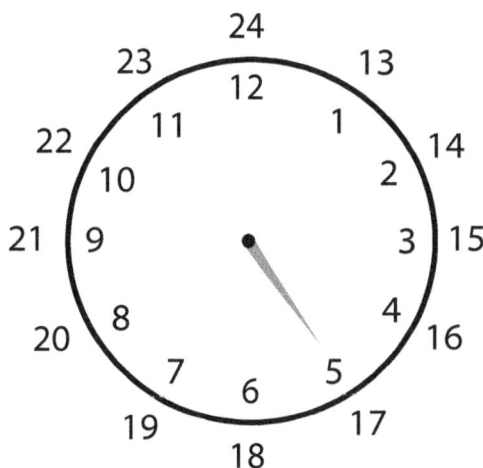

The relation of congruence is an equivalence relation.

(1) Reflexive $\quad x \equiv x$
example: $2 \equiv 2$

(2) Symmetric $\quad x \equiv y$ then $y \equiv x$
example: $3 \equiv 15$ then $15 \equiv 3$

(3) Transitive $\quad$ If $x \equiv y$ and $y \equiv z$ then $x \equiv z$
example: $3 \equiv 15$ and $15 \equiv 3$ then $3 \equiv 3$

References – The Structure of Algebra

[1] Bloch, Norman J., Abstract Algebra with Applications, Prentice-Hall, NJ, 1987.

[2] Devlin, Keith J., Sets, Functions and Logic: an introduction to abstract mathematics, 3rd Edition, Chapman & Hall/CRC, Fl. 2004.

[3] Fletcher, Peter, Hoyle, Hughes and Patty, C. Wayne, Foundations of Discrete Mathematics, PWS-Kent Publishing Company, Boston, MA, 1991.

[4] Gilbert, Linda and Gilbert, Jimmie, Elements of Modern Algebra, 7th Edition. Brooks/Cole, Cengage Learning, Belmont, CA, 2009.

[5] Nicodemi, Olympia E., Sutherland, Melissa A., and Towsley, Gary W., An Introduction to Abstract Algebra with Notes to the Future Teacher. Pearson/Prentice Hall, NJ. 2007

7
Modular Arithmetic

A Prelude to Modular Arithmetic

The Greatest Common Divisor

Let a and b be integers, not both zero. The greatest common divisor of a and b is the largest integer that divides both a and b.

Example Let a and b be integers, where $a = 30$ and $b = 66$.
The set of positive divisors of 30 is $\mathbf{S}_{30} = \{2, 3, 5, 6, 15, 30\}$.
The set of positive divisors of 66 is $\mathbf{S}_{66} = \{2, 3, 6, 11, 22, 33, 66\}$.

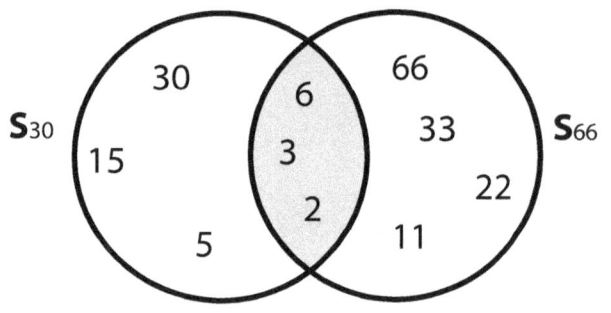

The set of common divisors is
$\mathbf{S}_{30} \cap \mathbf{S}_{66} = \{2, 3, 6\}$.

Therefore the greatest common divisor is 6, or $\gcd(30, 66) = 6$.

The Geometric View of gcd

A rectangular where side $a = 30$ and side $b = 66$ can be covered by 1,980 pieces of square tiles measuring 1 x 1, by 495 tiles 2 x 2, by 220 tiles 3 x 3, or by 55 tiles 6 x 6. In general, an a x b rectangular can be covered with square titles of the size c x c, if c is the common divisor of a and b.

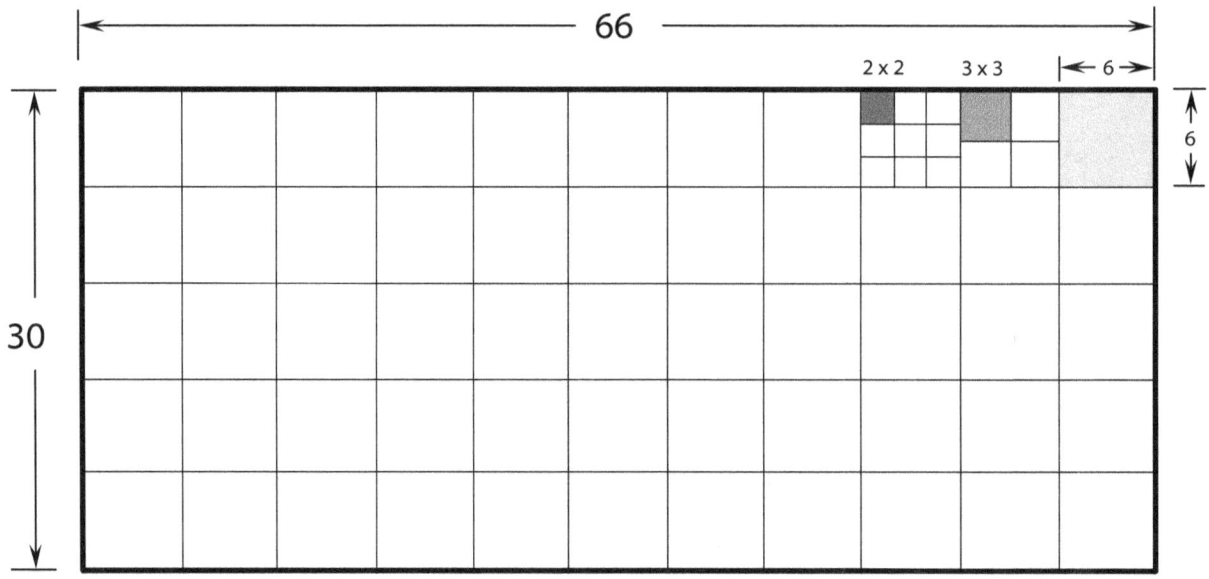

The largest square tile, 6 x 6, is the greatest common divisor of 30 and 36.

Euclidean Algorithm

The Euclidean Algorithm was stated by Euclid in his *Elements* of geometry over 2000 years ago. This algorithm has proved to be useful in a variety of mathematical contexts.

Finding the greatest common divisor of two integers

The Euclidean Algorithm is one of the most efficient ways to find the greatest common divisor of two integers. It is based on the following principle:

> Let a, b, q and r be integers, with $a \geq b$ and $a = b \cdot q + r$ (the remainder).
> Then $\gcd(a, b) = \gcd(b, r)$

In the division $a \div b$, a is the dividend and the smaller integer b is the divisor. The operation produces a quotient q and a remainder r. The gcd does not change if the dividend a is replaced by the remainder r. The replacement process can be done repeatedly until the smaller integer divides the larger one without remainder. The last nonzero remainder is the gcd of the original two numbers.

An iterative procedure

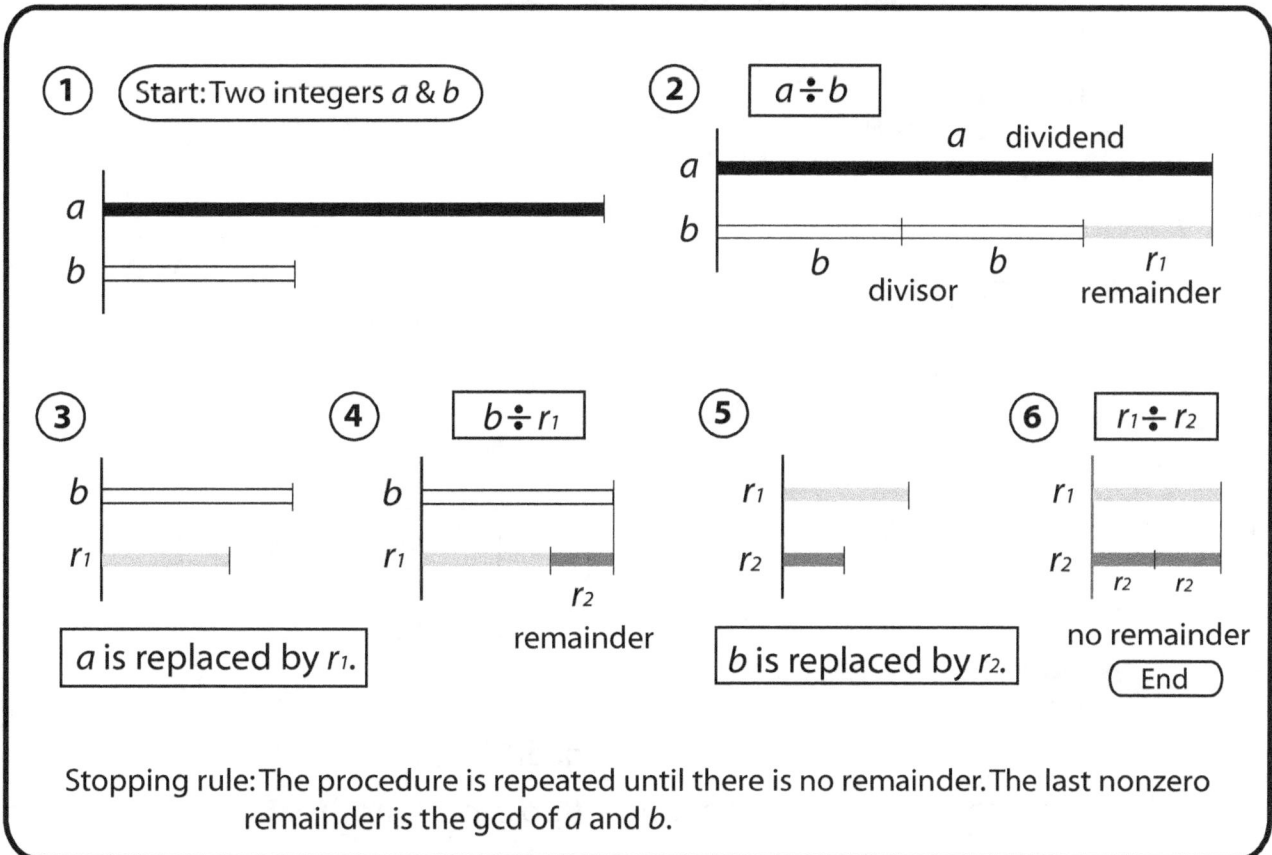

Example

A 60-inch and a 25-inch hardwood floor strips are cut up to shorter pieces with equal length. Any leftover part is not allowed. What is the longest possible length of each cut piece?

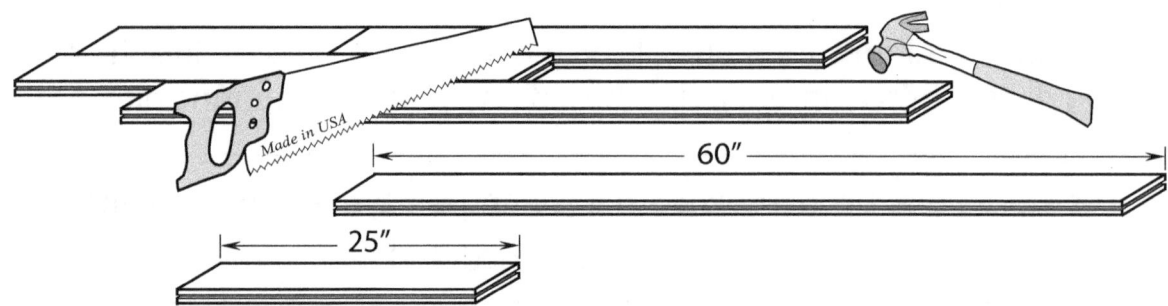

The longest possible length of each cut piece is the greatest common divisor of 60 and 25.

Step 1 Let $a = 60$ and $b = 25$

$\boxed{a \div b}$

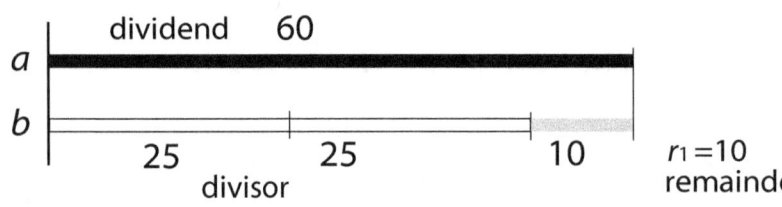

Step 2

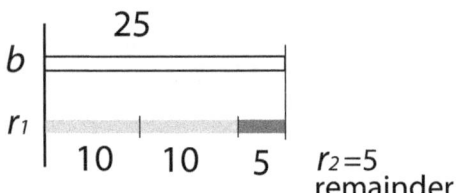

$\boxed{a \text{ is replaced by remainder } 10.}$

$\boxed{b \div r_1}$

Step 3

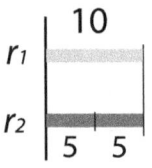

no remainder
(End)

$\boxed{b \text{ is replaced by remainder } 5.}$

$\boxed{r_1 \div r_2}$

$\gcd(60, 25) = 5$

The longest possible length is 5.

Example

Find gcd(76, 28) by Euclid's Algorithm

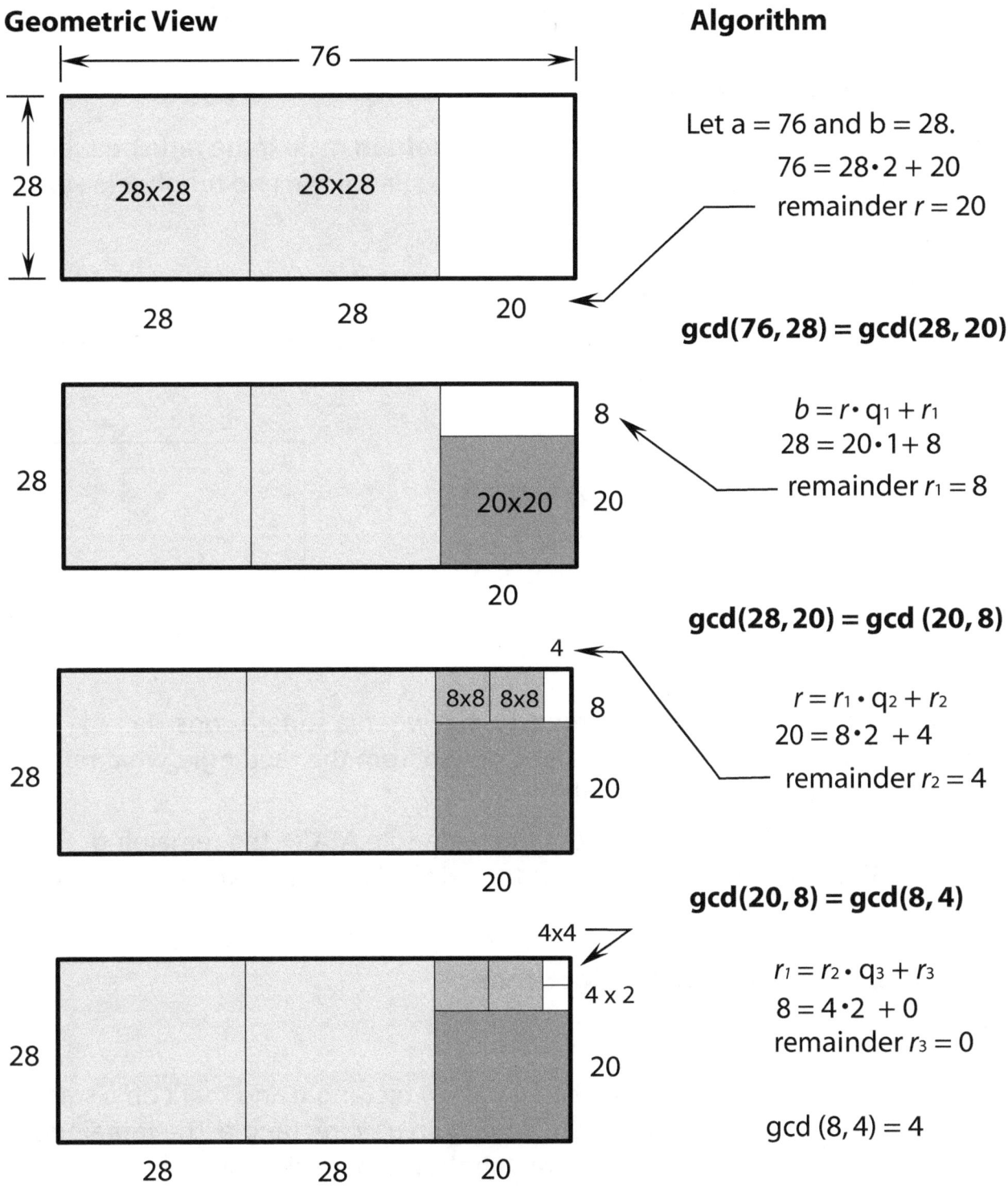

Geometric View

Algorithm

Let $a = 76$ and $b = 28$.
$76 = 28 \cdot 2 + 20$
remainder $r = 20$

gcd(76, 28) = gcd(28, 20)

$b = r \cdot q_1 + r_1$
$28 = 20 \cdot 1 + 8$
remainder $r_1 = 8$

gcd(28, 20) = gcd(20, 8)

$r = r_1 \cdot q_2 + r_2$
$20 = 8 \cdot 2 + 4$
remainder $r_2 = 4$

gcd(20, 8) = gcd(8, 4)

$r_1 = r_2 \cdot q_3 + r_3$
$8 = 4 \cdot 2 + 0$
remainder $r_3 = 0$

gcd (8, 4) = 4

Therefore, gcd(76, 28) = 4

Euclidean Algorithm and the Golden Ratio

In the Euclidean Algorithm, the division process is iterated until the remainder equals zero. What would happen if the remainder never equals zero?

The irrational golden ratio φ

Two numbers, a and b with $a > b$, are in the **golden ratio** if the ratio between them, a/b, is the same as the ratio between the sum of the two number $(a+b)$ and the larger of the two numbers, a.

Example: The Golden Rectangle

A golden rectangle is a rectangle where the ratio between the longer side and short side is in the golden ratio φ.

$$\text{Golden ratio } \varphi = \frac{a}{b} = \frac{a+b}{a}$$

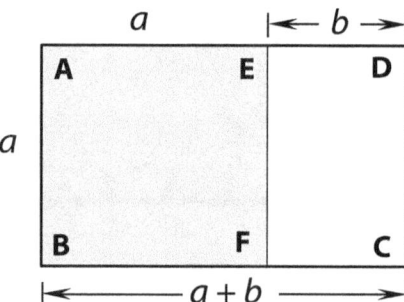

The Golden Rectangle has the property that when a square constructed on the shorter side of the rectangle is removed from the rectangle, what remains is another smaller Golden Rectangle.

If the square **ABFE** is removed from the rectangle **ABCD**, the remaining rectangle **EFCD** has the same shape as the original rectangle and is also a Golden Rectangle.

Find gcd(a, $a+b$) by Euclidean Algorithm:

Step ①

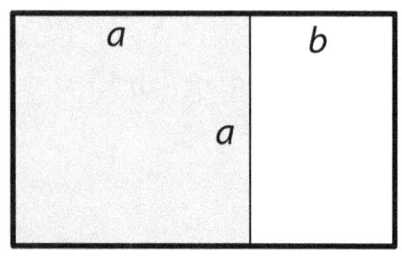

square $a \times a$ remainder

Draw an $a \times (a+b)$ rectangle and mark off a square on side a, leaving an $a \times b$ rectangle. The remainder, the $a \times b$ rectangle, will have the same shape as the original rectangle.

Step ②

Take the rectangle $a \times b$, and mark off a square on side b, leaving a $b \times c$ rectangle. The remainder keeps the same shape as the original rectangle.

Step ③   Step ④   Step ⑤ ...

At each step, the remainder - a smaller golden rectangle - is never equal to zero. The process never terminates, because the golden ratio φ is an **irrational number**,

$$\varphi = \frac{a}{b} = \frac{1+\sqrt{5}}{2} \cong 1.6180339887....$$

Congruence of Integers

In basic mathematics, a straight number line is commonly used in representing numbers and operations.

Integers system

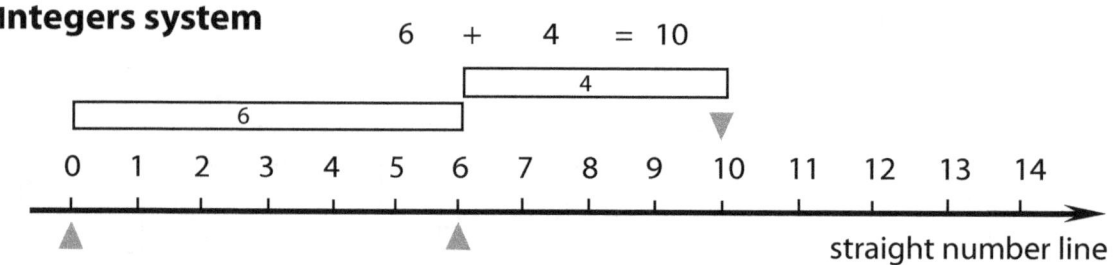

If an integer number line is wound in a circle of particular size, a new number system is created.

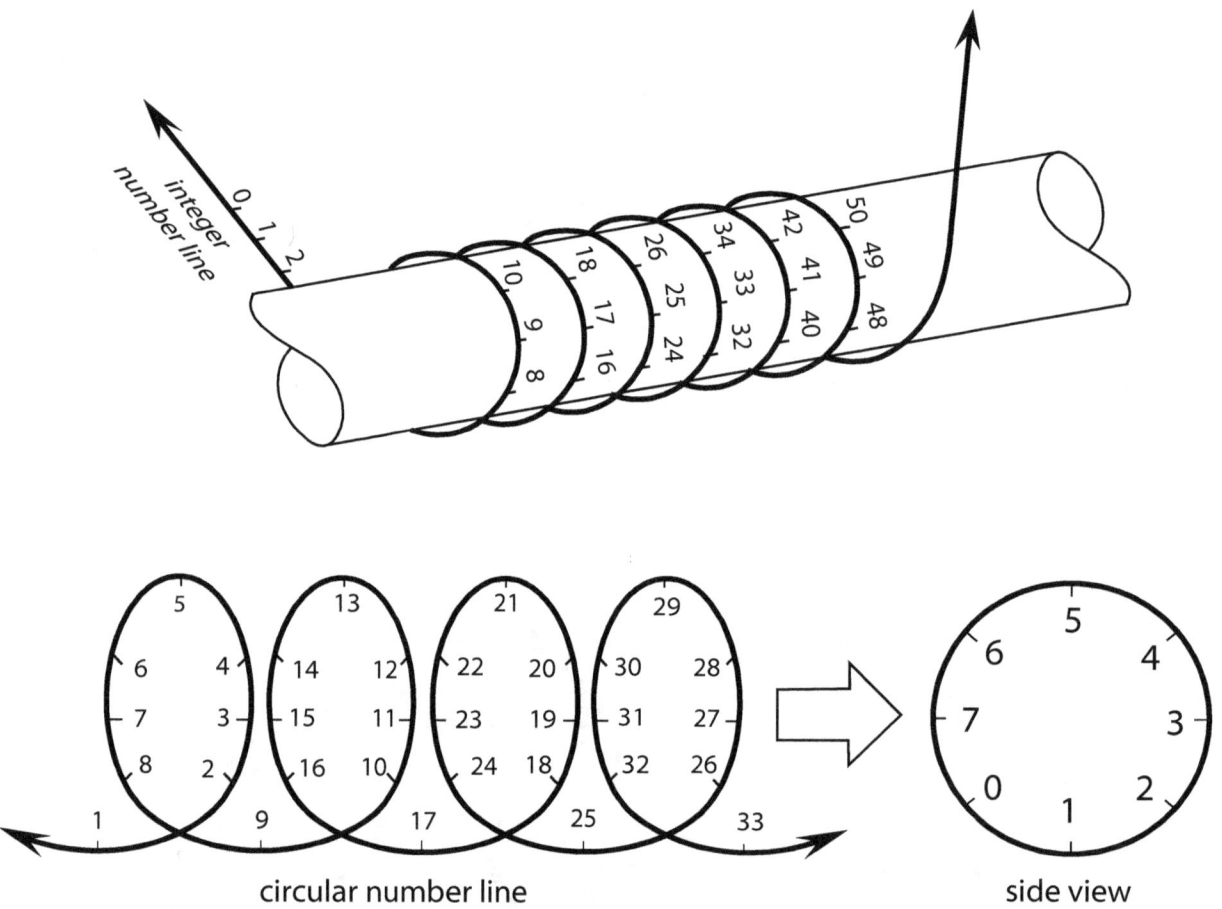

In this example, a circle with the size of 8 positions is used. The numbers at the same position of the circle are called **congruent**. The symbol "≡" is used to indicate the congruence. The size of the circle 8 is called the **modulus**.

0 ≡ 8 ≡ 16 ≡ 24 ≡ ...
1 ≡ 9 ≡ 17 ≡ 25 ≡ ...
2 ≡ 10 ≡ 18 ≡ 26 ≡ ...
3 ≡ 11 ≡ 19 ≡ 27 ≡ ...
4 ≡ 12 ≡ 20 ≡ 28 ≡ ...
5 ≡ 13 ≡ 21 ≡ 29 ≡ ...
6 ≡ 14 ≡ 22 ≡ 30 ≡ ...
7 ≡ 15 ≡ 23 ≡ 31 ≡ ...

There are 8 positions on the circle. Numbers 8, 16, and 24 are at the same position as 0 on the circle. Hence, they are congruent to 0. Likewise, numbers 9, 17 and 25 are congruent to 1, and so on.

Carl Friedrich Gauss
1777 - 1855
The Prince of Mathematics

If the difference of two integers a and b is divided by an integer n then a and b are **congruent** with respect to n. The integer n is called the **modulus**.

--- Gauss, Disquisitiones Arithmeticae, 1801

Gauss's statement can be rewritten as follows:

> Let n be a positive integer, for integers a and b
>
> if $(a - b)/n = k$ or $a - b = nk$ for some integer k
>
> then a is congruent to b, $a \equiv b$ (modulo n)

The relation of congruence, ≡, is an equivalence relation. It indicates the two objects are alike in some respect, but not exactly the same as equal "=".

Congruence and Remainder

Let *a* and *b* be any two integers and *n* be any positive integer.

If *a* and *b* are divided by *n* and leave the **same remainder** *r*,

$$a = n q_1 + r \quad \text{and} \quad b = n q_2 + r$$

Subtracting *b* from *a*, we have

$$a - b = (n q_1 + r) - (n q_2 + r) = n(q_1 - q_2), \; n \text{ divides } a - b$$

Therefore *a* is congruent to *b*, $a \equiv b \pmod{n}$

> We say that two integers *a* and *b* are congruent modulo *n* if they leave the same remainder when divided by *n*.

Modular Arithmetic in Everyday Life

Other than the example in Chapter 6, there are more examples in daily life like the 7-day week, 12-month year, and computer's binary system, spirograph, etc.

A seven-day cycle

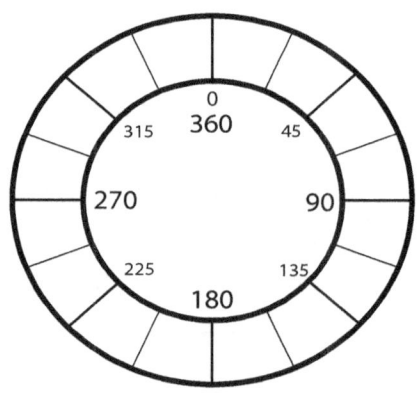

A 360-degree circle

Modular Arithmetic in Everyday Life (continued)

Gears

Zodiac

Spirograph

Computers, Cellular Phones and Digital Devices

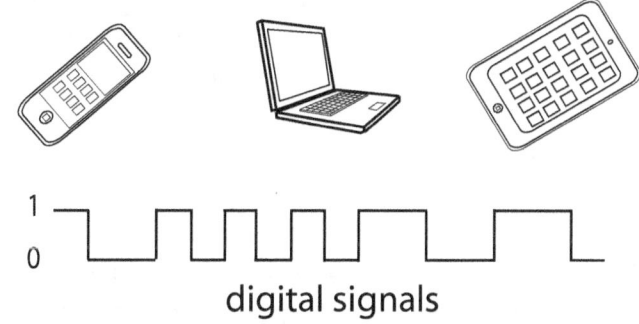

digital signals

The Clave Rhythm

The clave is a rhythmic pattern used as a tool for the temporal organization of Afro-Cuban music. Its structure can be understood in terms of cross-rhythmic ratios, three-to-two or two-to-three.

©2015 Chih Yang

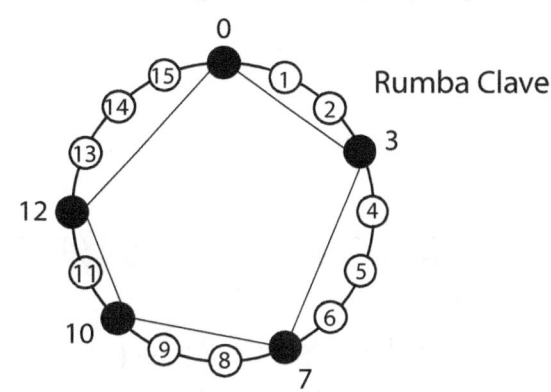

Rumba Clave

sixteen-step cycle

The Operations of Modular Numbers

Modular numbers can be added or multiplied the same way as equality "=" can.

Example: Addition for modulo 8 numbers,
$$6 + 4 \equiv 10 \equiv 2 \text{ (modulo 8)}$$

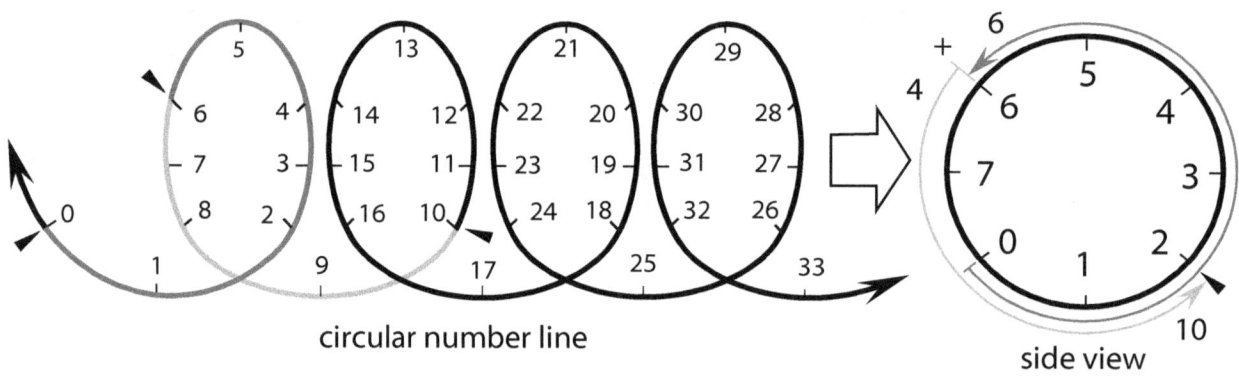

circular number line   side view

Properties of Addition and Multiplication

If $a \equiv b$ (modulo n) and k is an integer, then

$$a + k \equiv b + k \text{ (modulo } n\text{), and}$$

$$ak \equiv bk \text{ (modulo } n\text{)}$$

Example:

$$8 + 11 + 37 + 59 \equiv 1 + 4 + 2 + 3 \equiv 10 \equiv 3 \text{ (modulo 7)}$$

$$(5)(13)(39)(109) \equiv (5)(4)(3)(1) \equiv 60 \equiv 6 \text{ (modulo 9)}$$

$$41^{15} \equiv 2^{15} \equiv (2^4)^3 2^3 \equiv (16)^3 2^3 \equiv (3)^3 2^3 \equiv (27) 8$$
$$\equiv (1)(8) \equiv 8 \text{ (modulo 13)}$$

Applications

Example 1 UPC (Universal Product Codes) Symbols

UPC symbols consist of 12 digits, $a_1, a_2 \cdots a_{12}$. For error control, the last digit a_{12} is designated as the check bit and a weighting vector of 11-tuple (3, 1, 3, 1, 3, 1, 3, 1, 3, 1, 3) is used to find it.

The last digit a_{12} can be computed as

$$a_{12} = -(0, 3, 6, 0, 0, 0, 2, 9, 1, 4, 5) \cdot (3, 1, 3, 1, 3, 1, 3, 1, 3, 1, 3)$$
$$= -48 \equiv 2 \text{ (modulo 10)}$$

In the early 1990s, Sprague Ackley conducted researches sponsored by the US Department of Agriculture to track the behavior of bees by attaching a barcode to the back of bees. It led to the development of several new decoding techniques. (Hamblen, 2014)

Example 2 Bank Identification Numbers

Every bank has its own identification numbers. To avoid errors during data transactions, the identification numbers are coded with a check digit.

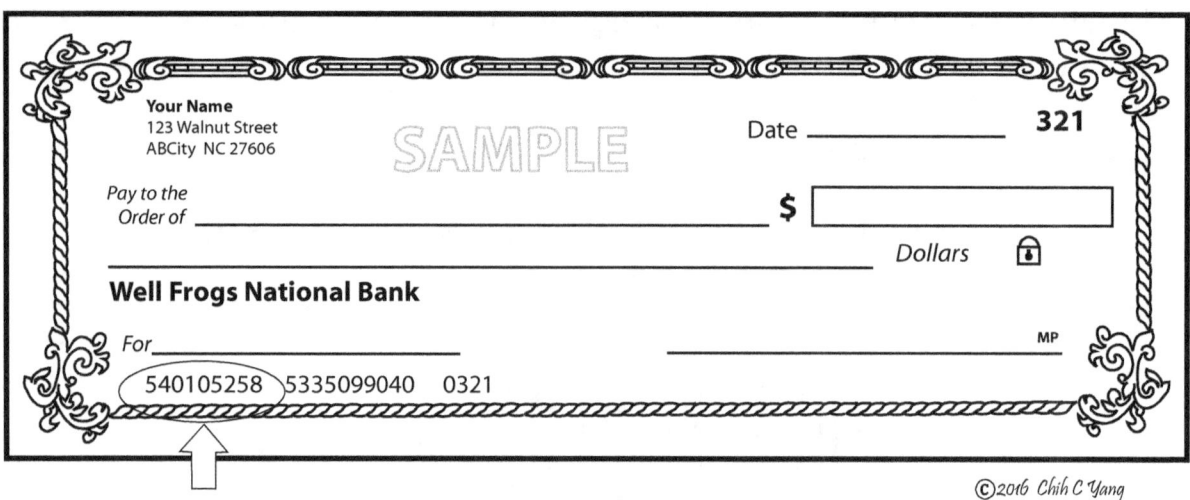

Bank identification number

Bank identification numbers consist of 8 digits, $b_1\, b_2 \ldots b_8$, and a check digit, b_9, for error detection.

The check digit b_9 can be computed with the weighting vector $(7, 3, 9, 7, 3, 9, 7, 3)$.

$$(b_1, b_2, \ldots, b_8) \cdot (7, 3, 9, 7, 3, 9, 7, 3) \equiv b_9 \text{ (modulo 10)}$$

On the sample check, the check digit of bank ID # 54010525 is

$$\begin{aligned} b_9 &= (5, 4, 0, 1, 0, 5, 2, 5) \cdot (7, 3, 9, 7, 3, 9, 7, 3) \\ &= 35 + 12 + 0 + 7 + 0 + 45 + 14 + 15 \\ &= 128 \equiv 8 \text{ (modulo 10)} \end{aligned}$$

The complete ID number is shown as 540105258.

If the ID was incorrectly keyed in as 540106258, the check digit would be calculated as 7 rather than 8. Thus, the error is detected.

Example 3 Applications in Bookkeeping

In accounting, there is a rule to check for arithmetic errors based on the following statement.

> A positive integer is divisible by 9 if and only if the sum of its digits is divisible by 9.

Example:

$7154358245964 \div 9 = 794928693996$

7154358245964 is divisible by 9.

$7 + 1 + 5 + 4 + 3 + 5 + 8 + 2 + 4 + 5 + 9 + 6 + 4 = 63$

The sum of its digits 63 is divisible by 9.

Explanation:

> Any integer z could be expressed in the form as:
>
> $$z = a_0 + a_1 \cdot 10 + a_2 \cdot 10^2 + \ldots + a_n \cdot 10^n$$
>
> $\because$ $10 \equiv 1$ (modulo 9)
>
> $$z \equiv a_0 + a_1 \cdot 1 + a_2 \cdot 1^2 + \ldots + a_n \cdot 1^n \text{ (modulo 9)}$$
>
> $\therefore$ $z \equiv a_0 + a_1 + a_2 + \ldots + a_n$ (modulo 9)
>
> If z is divisible by 9 then $z \equiv 0$ (modulo 9)
>
> and $z \equiv a_0 + a_1 + a_2 + \ldots + a_n \equiv 0$ (modulo 9)
>
> Therefore the sum of its digits is divisible by 9.

Casting Out 9s

One application from the above statement is the "casting out 9s" method.

Casting Out 9s method
(1) delete all 9s or any group of digits whose sum is 9
(2) add up all the remaining digits and find its congruence of 9

Example: Testing Arithmetic Errors by Casting out 9s

Is the statement (7654321 + 35791) x 12345 = 94934432640 correct?

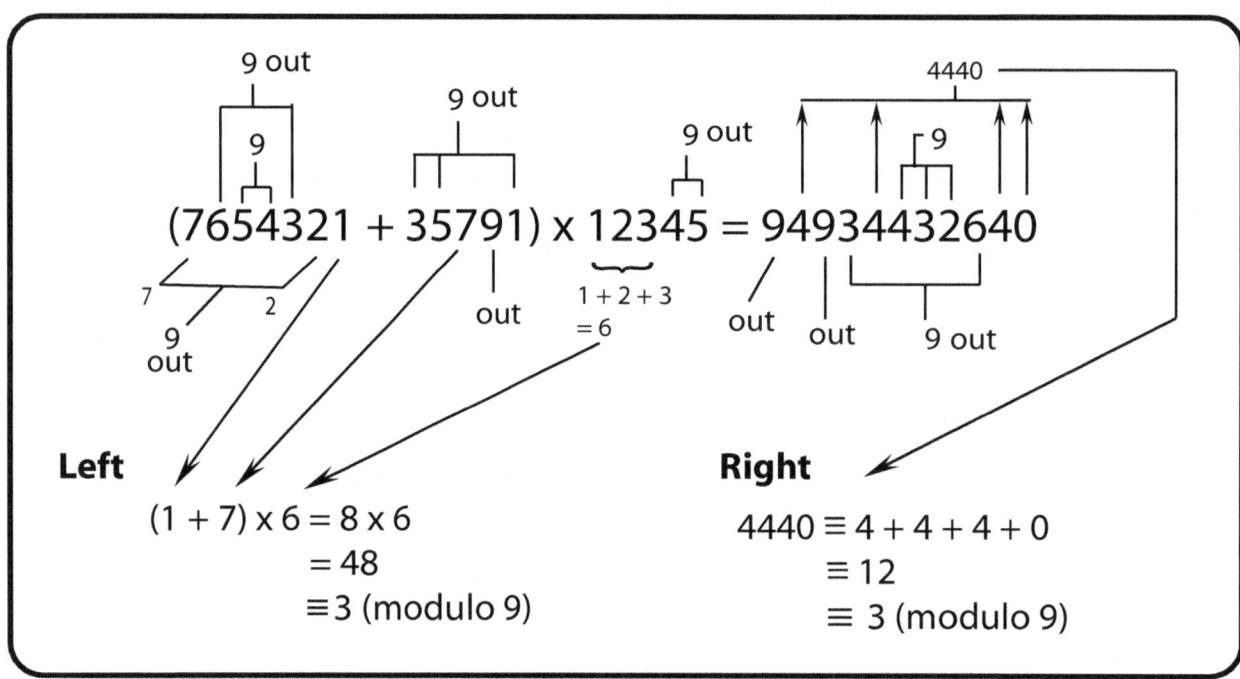

Both sides have the same remainder. Thus, it passed the test of casting out 9s. No errors were detected. If the two sides were different, then the computation was not correct.

This method is not infallible in detecting errors. If a transposition occurred, an error would not be detected.

Example 4 Detecting Transposition Errors

At a fair event there is a discrepancy between received cash and book records.

$$\$12{,}124.79 - \$11{,}764.79 = \$360.00$$

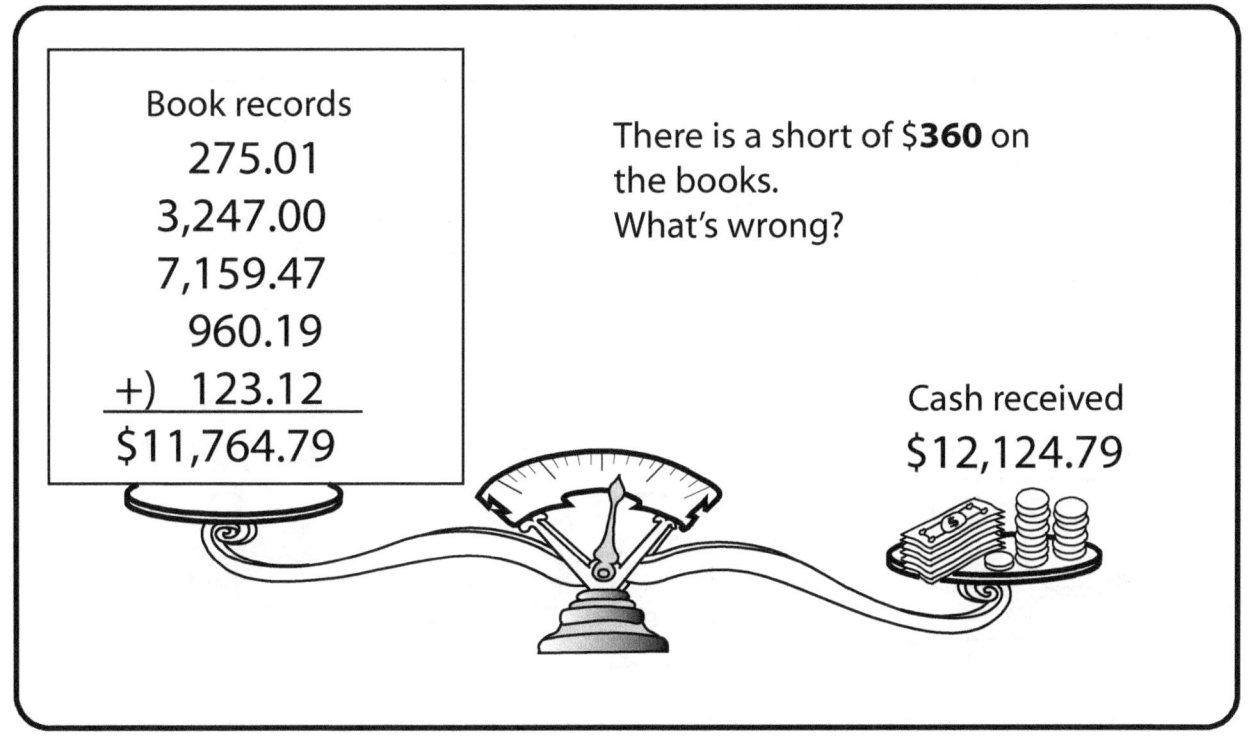

Book records
275.01
3,247.00
7,159.47
960.19
+) 123.12
$11,764.79

There is a short of **$360** on the books.
What's wrong?

Cash received
$12,124.79

Two hints can assist in finding the error:

> (1) If the difference is divisible by 9. There could be a transposition error.
> (2) Since the difference of $360, has 3 digits and the first digit was 0, the transposition could have occurred at the 2nd and 3rd digits.

Check each item with the 2nd - 3rd digit transposition as follows:

⇊ ⇊
275.01 - 725.01 = -450
3,247.00 - 3,427.00 = -180
7,159.47 - 7,519.47 = **-360** ← the discrepancy
960.19 - 690.19 = 270
123.12 - 213.12 = -90

The number 7,159.47 must be wrong. The correct amount should be 7,519.47.

Change the number and add them up again:

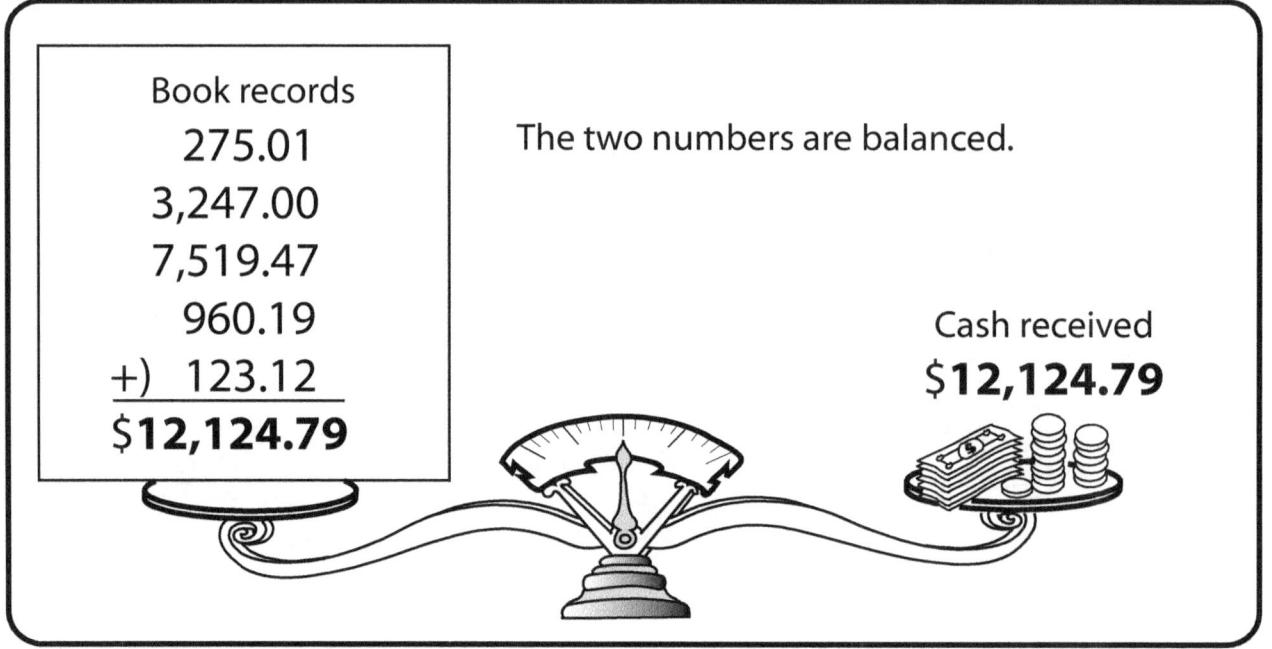

References – Modular Arithmetic

[1] Bloch, Norman J., Abstract Algebra with Applications, Prentice-Hall, NJ, 1987.

[2] Gallian, Joseph A., Contemporary Abstract Algebra, 8th Edition, Brooks/Cole, Cengage Learning, Blemont, CA, 2013.

[3] Gilbert, Linda and Gilbert, Jimmie, Elements of Modern Algebra, 7th Edition. Brooks/Cole, Cengage Learning, Belmont, CA, 2009.

[4] Hamblen, Matt, The UPC Bar code arrived 40 years ago; now, they're ubiquitous, Computerworld, Framingham, Mass, June 23, 2014.

[5] Nicodemi, Olympia E., Sutherland, Melissa A., and Towsley, Gary W., An Introduction to Abstract Algebra with Notes to the Future Teacher.Pearson/Prentice Hall, NJ. 2007

[6] Rotman, Joseph J., A First Course in Abstract Algebra with Applications, 3rd Edition, Pearson Education, NJ, 2006.

[7] Stewart, Ian, Concepts of Modern Mathematics, Dover Publications. 1995

[8] Toussaint, Godfried, The Euclidean Algorithm Generates Traditional Musical Rhythms, Proceedings of BRIDGES: Mathematical Connection in Art, Music, and Science. Banff, Alberta, Canada, July 31 – August 3, 2005

8
Chinese Counting

Counting

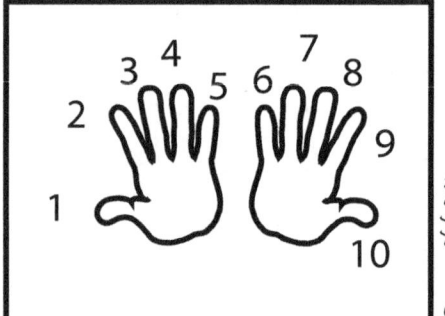
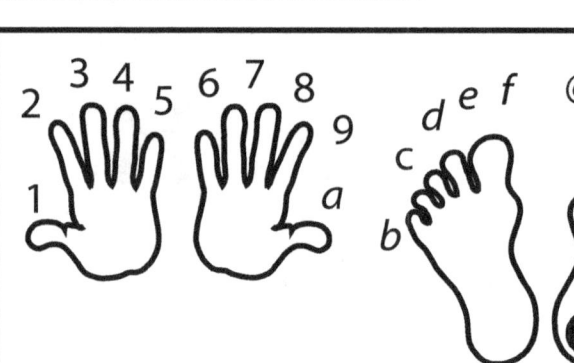

Tar Heel!

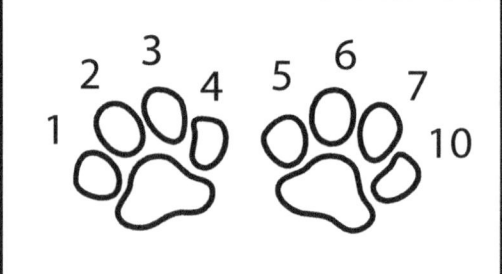

Chinese Remainder Counting

There is a story in the ancient Chinese mathematics book, Sun Tsu Suan Ching (孫子算經, ~ 400 AD). It tells how a Chinese general Han Shin (韓信, ~196 BC) counted his troops.

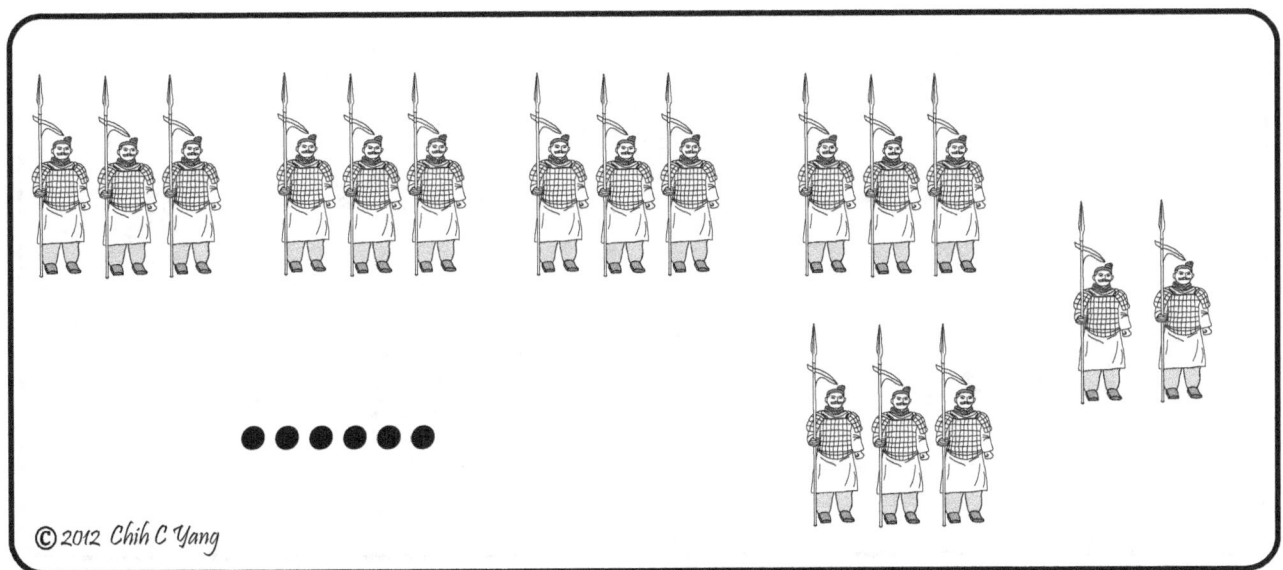

In a large field, he instructed them to form groups of three, and counted the remainder.

Chinese Remainder Counting (continued)

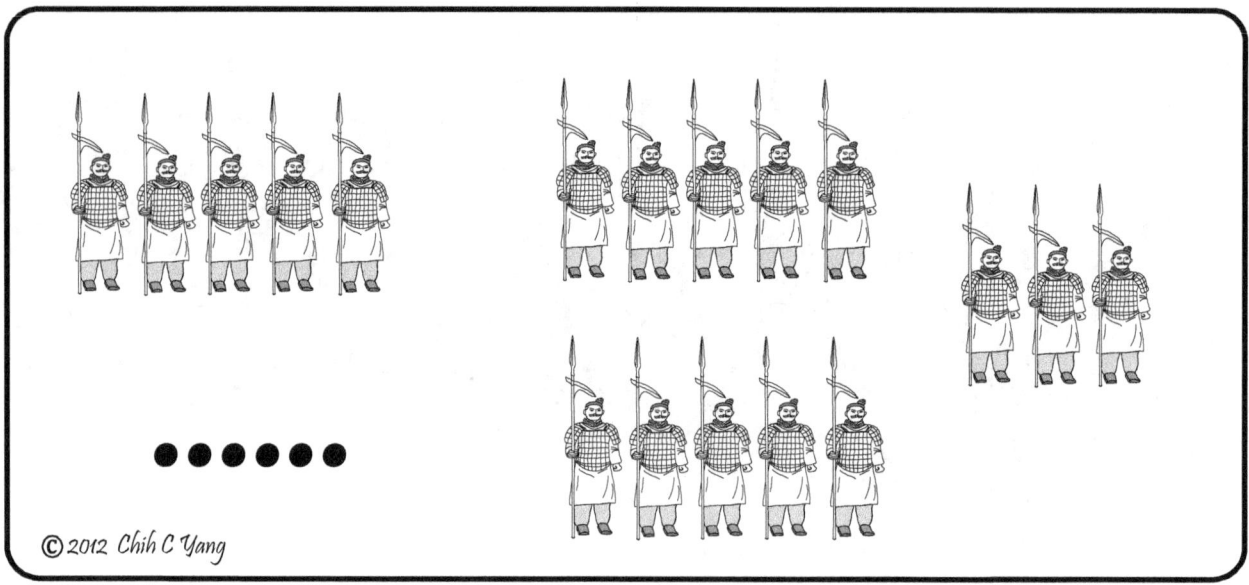

He then instructed the soldiers to reassemble into groups of five and counted the remainder.

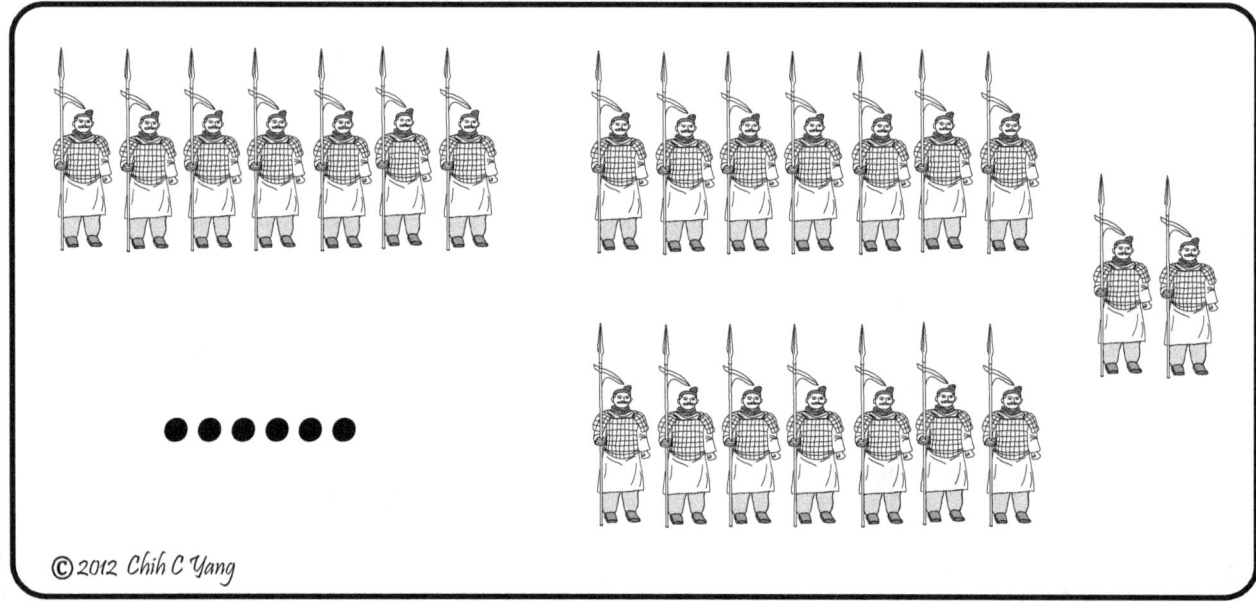

He repeated the process with groups of seven, then and was then able to figure out the exact number of troops.

Why did he count using this method?

The number of soldiers, directly correlated with fighting capacity, has to be encrypted.

How did he decrypt it?

Step 1 Line up the remainders as shown below and prepare three different sized blocks: 3-unit, 5-unit, and 7-unit.

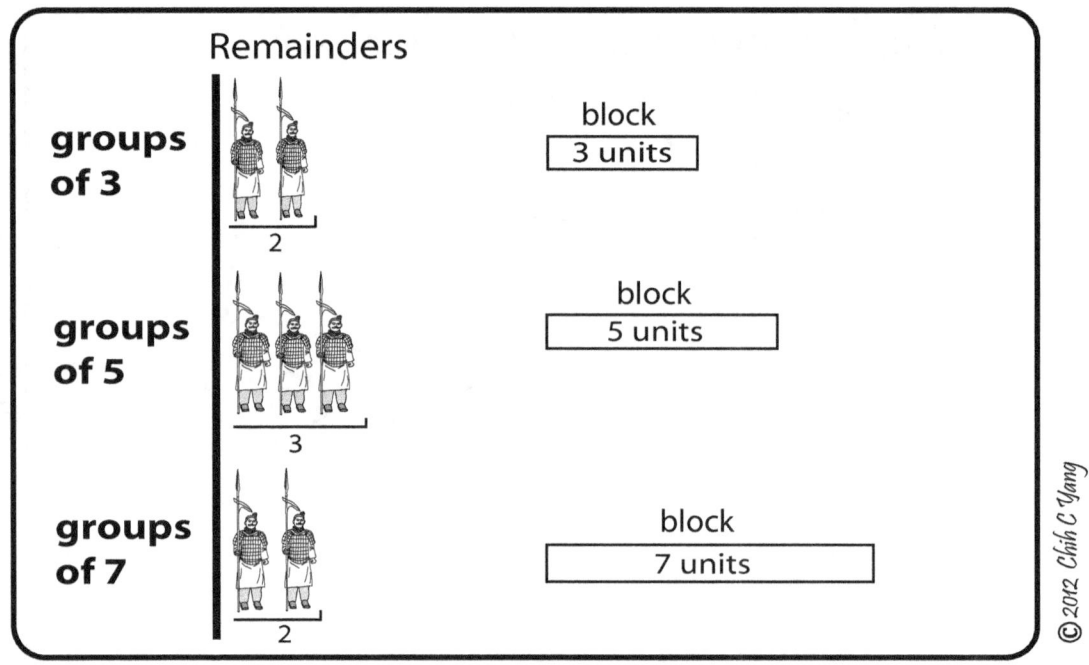

Step 2 Line up the blocks

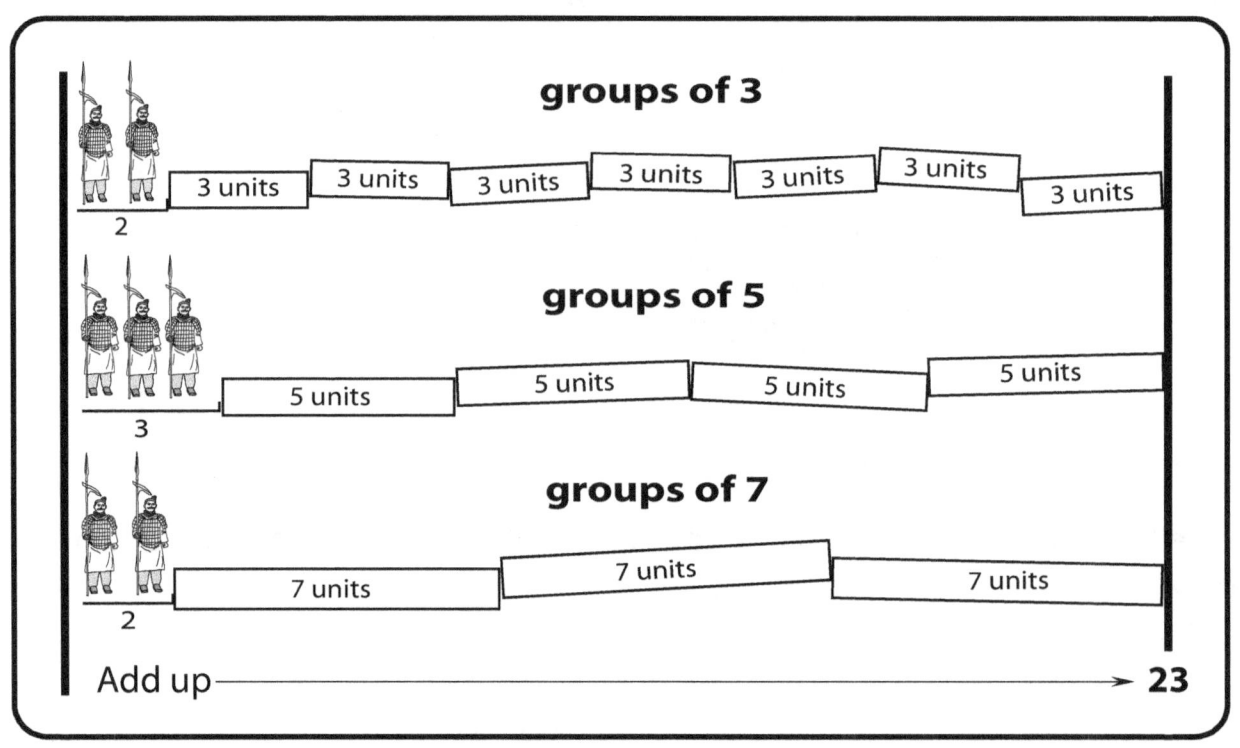

The number is 23!

Mathematical Description

This problem can be described in the form of modular numbers.

Let x be the number of soldiers, then

$$\begin{cases} x \text{ divided by 3 with a remainder of 2} \\ x \text{ divided by 5 with a remainder of 3} \\ x \text{ divided by 7 with a remainder of 2} \end{cases}$$

Written as a system of equations: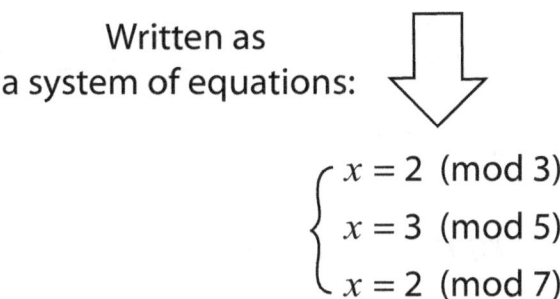

$$\begin{cases} x = 2 \ (\text{mod } 3) \\ x = 3 \ (\text{mod } 5) \\ x = 2 \ (\text{mod } 7) \end{cases}$$

There are two methods to find the solution.

(1) Brute-Force Solution

Each remainder has its own congruence class* [a] (mod n):

$$x = 2 \in [2] \ (\text{mod } 3) = \{ ..., 2, 5, 8, 11, 14, 17, 20, \widehat{23}, ...\}$$
$$x = 3 \in [3] \ (\text{mod } 5) = \{ ..., 3, 8, 13, 18, \widehat{23}, ...\}$$
$$x = 2 \in [2] \ (\text{mod } 7) = \{ ..., 2, 9, 16, \widehat{23}, ...\}$$

The intersection of [2] (mod 3) ∩ [3] (mod 5) ∩ [2] (mod 7) = {23, 128, 233, ...}

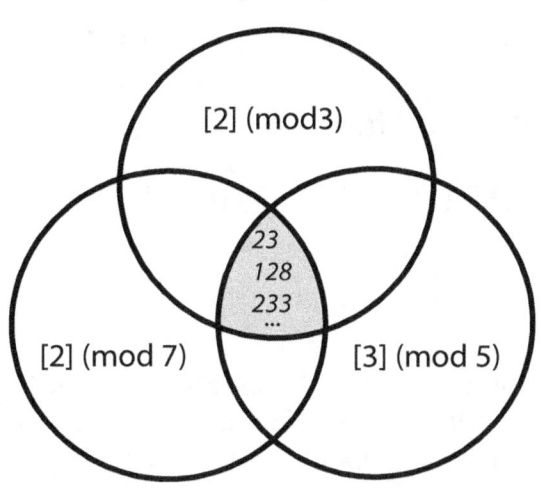

Number 23 is *the least common* number.
Thus $x = 23$

(2) General Solution

Chinese Remainder Theorem

> For a system of n equations, let $m_1, m_2, \cdots, m_n$ be pairwise *coprime*.
> $$x \equiv R_1 \pmod{m_1}$$
> $$x \equiv R_2 \pmod{m_2}$$
> $$\cdots$$
> $$x \equiv R_n \pmod{m_n}$$
>
> define $M = m_1 m_2 \cdots m_n$, $b_i = M/m_i$ and b_i^{-1} be the inverse of $b_i \pmod{m_i}$.
>
> The system has a solution for x modulo M.
>
> $$x = \sum_{i=1}^{n} R_i \, b_i \, b_i^{-1} \pmod{M}$$

(A) $m_1, m_2, ...,$ and m_n are pairwise coprime. Any two numbers m_i and m_j are relatively prime, when $i \neq j$.

(B) b_i^{-1} is the inverse of $b_i \pmod{m_i}$. In other word, $b_i (b_i^{-1}) \equiv 1 \pmod{m_i}$

When $R_1 = 2, R_2 = 3, R_3 = 2, m_1 = 3, m_2 = 5,$ and $m_3 = 7$,
(i) For $b_1 = M/m_i$, $b_1 = (m_1 m_2 m_3)/m_1 = m_2 m_3 = 5 \cdot 7 = 35$
Since $b_1(b_1^{-1}) \equiv 1 \pmod{m_1}$, $35(b_1^{-1}) \equiv 1 \pmod{3}$,
then $b_1^{-1} = 2$.

(ii) Similarly, we found $b_2 = 21, b_2^{-1} = 1, b_3 = 15$ and $b_3^{-1} = 1$

(iii) Substitute the above numbers into the following equation,

$$x = R_1 b_1 b_1^{-1} + R_2 b_2 b_2^{-1} + R_3 b_3 b_3^{-1} \pmod{M}$$

$$x = (2)(35)(2) + (3)(21)(1) + (2)(15)(1) = 233 \equiv 23 \pmod{105}$$

Thus $\boxed{x = 23}$

Note: A congruence class $[a] \pmod{n}$ is a set of numbers in which all numbers have the same remainder a when they are divided by modular n. In the case of congruence class $[2] \pmod{3}$, all elements of the set have the remainder of 2 when they are divided by 3.

Pirates of the Carolinas

A band of 17 pirates wanted to divided their loot of gold coins evenly. When they divided the coins into equal piles, 5 coins were left over. In a brawl over who should get the extra coins, two of the pirates were slain. Again, the remaining 15 pirates tried to divide the loot equally, 3 coins remained. Another pirate was killed in another fight over the extra coins. Finally the last 14 pirates divided the loot equally among themselves.

What was the least number of coins that the pirates could have distributed?

(See appendix for answers)

Applications

Example 1 Computer Precision Arithmetic

The Chinese Remainder Theorem (**CRT**) is used in parallel computations. It speeds up calculations of huge integers with hundreds of digits.

Every computer has a limit on the number of integers that can be used in a calculation, called the word size. Calculations with integers larger than the word size are very time-consuming. In such cases, the Chinese Remainder Theorem offers a faster approach.

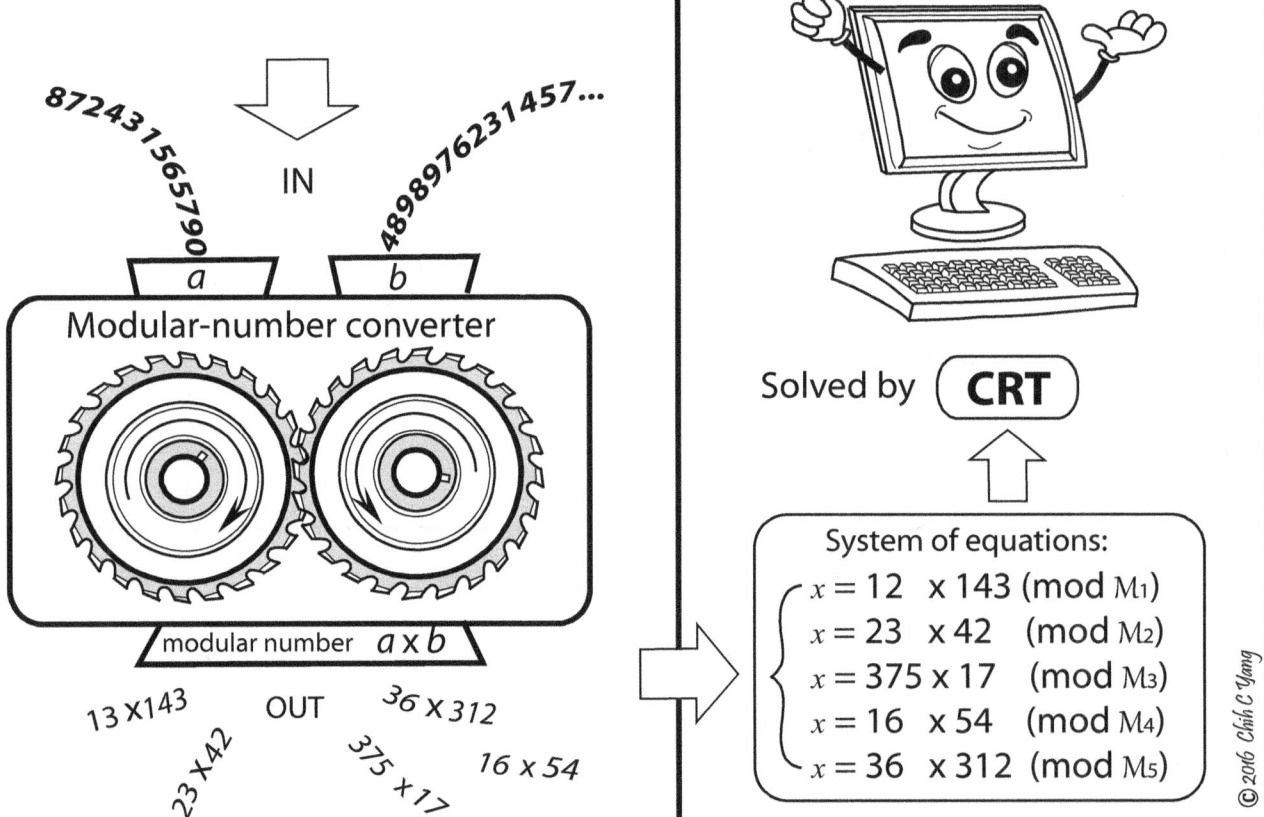

The two large numbers are converted into sets of smaller modular numbers which then form a system of equations.

Example: Find $x = 4567 \times 8976 = ?$ using the Chinese Remainder Theorem

Step 1 Convert both 4567 and 8976 into modular numbers.

To keep the remainders in the single or double digits, use numbers 89, 95, 97, 98 and 99 as moduli in this example.

$$4567 \equiv 28 \pmod{89} \quad\quad 8976 \equiv 76 \pmod{89}$$
$$4567 \equiv 7 \pmod{95} \quad\quad 8976 \equiv 46 \pmod{95}$$
$$4567 \equiv 8 \pmod{97} \quad\quad 8976 \equiv 52 \pmod{97}$$
$$4567 \equiv 59 \pmod{98} \quad\quad 8976 \equiv 58 \pmod{98}$$
$$4567 \equiv 13 \pmod{99} \quad\quad 8976 \equiv 66 \pmod{99}$$

Step 2 Rewrite 4567×8976 as a product of modular numbers.

$$4567 \times 8976 \equiv 28 \times 76 \equiv 81 \pmod{89}$$
$$4567 \times 8976 \equiv 7 \times 46 \equiv 37 \pmod{95}$$
$$4567 \times 8976 \equiv 8 \times 52 \equiv 28 \pmod{97}$$
$$4567 \times 8976 \equiv 59 \times 58 \equiv 90 \pmod{98}$$
$$4567 \times 8976 \equiv 13 \times 66 \equiv 66 \pmod{99}$$

Step 3 Solve the following system of congruences:

$$\begin{cases} x \equiv 81 \pmod{89} \\ x \equiv 37 \pmod{95} \\ x \equiv 28 \pmod{97} \\ x \equiv 90 \pmod{98} \\ x \equiv 66 \pmod{99} \end{cases}$$

Using the Chinese Remainder Theorem,

we find $x = 40,993,392$.

Note: Two small numbers, $a = 4567$ and $b = 8976$, are chosen for demonstration purposes.

Example 2 - Cryptography

Cryptography is a technique to design codes for data and information security. There are two kinds of cryptographic systems.

(1) Private-Key System

A system in which the same secret key is used to encrypt and decrypt messages.

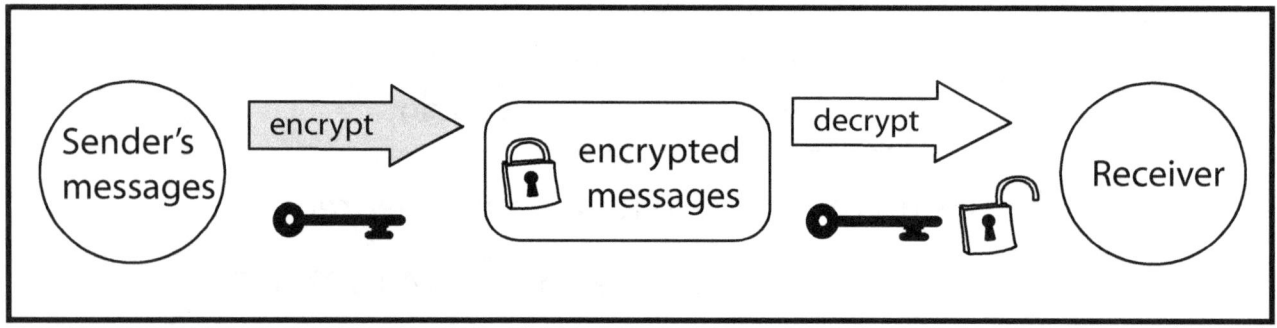

(2) Public-Key System

A system that uses two different keys - a private key and a public key.

RSA Public-key Cryptosystem

The RSA cryptosystem is the most widely used public-key cryptography. It is used for securing web traffic, e-mail, and credit-card payment systems. The system is based on the difficulty of factoring the product of two large prime numbers.

How does RSA cryptosystem work?

Step 1 Make a pair of keys

Two *large* prime numbers are required to make the keys. For demonstration purposes, two small prime numbers $p = 23$ and $q = 43$ are chosen. The two keys, derived from p and q, are shown below:

Public key

encrypt equation $f(x) = x^e \pmod{m}$

$m = pq = 23 \times 43 = 989$ and $e = 221$,

$$f(x) = x^{221} \pmod{989}$$

(See appendix for details)

(to be made public)

Private key

decrypt equation $g(x) = x^d \pmod{m}$

$d = 485, \quad g(x) = x^{485} \pmod{989}$

(to be kept secret)

Step 2 Convert alphabet letters to numbers

Conversion Table: Associate the letters of the alphabet with integers.

Alphabet	a	b	c	d	e	f	g	h	i	j	k	l	m	n
Number	00	01	02	03	04	05	06	07	08	09	10	11	12	13
Alphabet	o	p	q	r	s	t	u	v	w	x	y	z	blank	
Number	14	15	16	17	18	19	20	21	22	23	24	25	26	

Example: A plaintext message "buddy i found a bone" is converted to numbers and arranged into 3-digit groups.

Plaintext	b u d d y	i	f o u n d	a	b o n e
Number	01 20 03 03 24	26 08 26	05 14 20 13 03	26 00 26	01 14 13 04 26
3-digit groups	012 003 032 426 082 605 142 013 032 600 260 114 130 426				

Step 3 Encrypt the plaintext message

The 3-digit groups are translated by the public key, a mapping f, into ciphertext as follows:

$$\text{Mapping } f: x \rightarrow f(x)$$

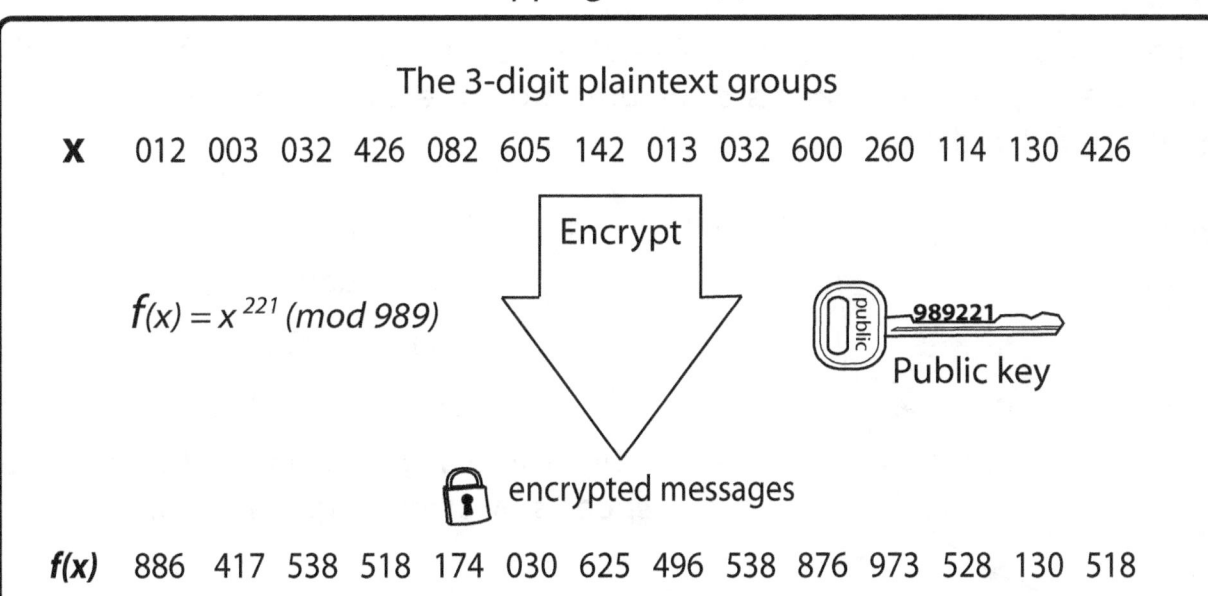

The 3-digit plaintext groups

x 012 003 032 426 082 605 142 013 032 600 260 114 130 426

Encrypt

$f(x) = x^{221} \;(mod\; 989)$

Public key 989221

encrypted messages

f(x) 886 417 538 518 174 030 625 496 538 876 973 528 130 518

Step 4 Decrypt the cipher

Step 4 is the inverse of steps 2 and 3. The private key, mapping g, is used to decrypt the ciphertext. In this example, $g(x) = x^{485} \pmod{989}$.

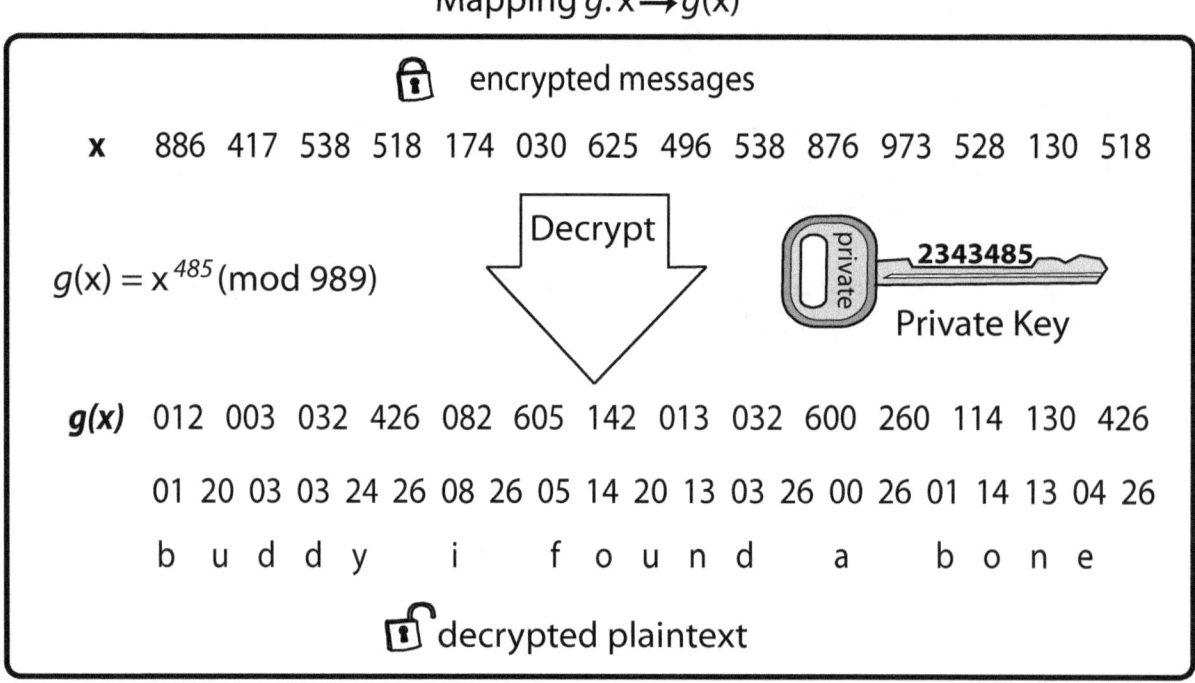

A Shortcoming of Public-Key Systems

Public-key systems work very well and are extremely secure, but can consume a large amount of computing power and are significantly slower than private-key systems.

In practice, the RSA system requires two large prime numbers, p and q. This generates very large exponents, e and d, for both the encryption and decryption equations, $f(x)$ and $g(x)$.

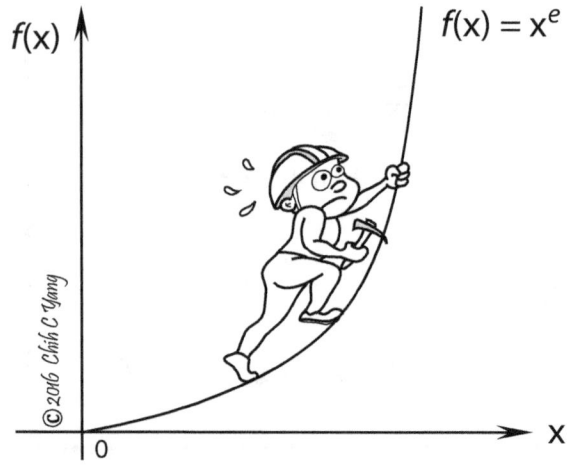

For equations,

$$f(x) = x^e \pmod{m}, \text{ and}$$

$$g(x) = x^d \pmod{m}$$

when e and d are very large, the computation speed slows down dramatically.

Increasing the Efficiency of RSA Implementations with Chinese Remainder Theorem (CRT)

Most implementations of RSA use the Chinese Remainder Theorem for signing HTTPS certificates and during decryption. This method is much faster than the standard RSA algorithm alone without CRT.

How does it work?

Standard RSA algorithm without CRT

In standard CRT operations, the calculations are performed directly with equations $f(x)$ and $g(x)$. Let $x_1 = a$, and $x_2 = b$

Then, Public key $f(x1) = a^e \pmod{m}$

Private key $g(x2) = b^d \pmod{m}$

Slow

Implementations with CRT

Step 1

Replace the moduli m, in both public and private keys with smaller numbers p and q and rewrite the equations as a system of equations:

$f(x_1) = x_1^e \pmod{m}$, where $m = pq$ $\Rightarrow$ $\begin{cases} f(x_1) = x_1^e \pmod{p} \\ f(x_1) = x_1^e \pmod{q} \end{cases}$

$g(x_2) = x_2^d \pmod{m}$, where $m = pq$ $\Rightarrow$ $\begin{cases} g(x_2) = x_2^d \pmod{p} \\ g(x_2) = x_2^d \pmod{q} \end{cases}$

Step 2

Use the system of equations as keys, let $x_1 = a$ and $x_2 = b$, then,

$\begin{cases} f(x_1) = a^e \pmod{p} = \mathbf{A_1} \\ f(x_1) = a^e \pmod{q} = \mathbf{A_2} \end{cases}$ $\Rightarrow$ Solved by CRT $\Rightarrow$ $f(x_1) = \mathbf{A} \pmod{m}$

$\begin{cases} g(x_2) = b^d \pmod{p} = \mathbf{B_1} \\ g(x_2) = b^d \pmod{q} = \mathbf{B_2} \end{cases}$ $\Rightarrow$ $g(x_2) = \mathbf{B} \pmod{m}$

Fast

Example

Suppose that in an RSA public key cryptosystem, $p = 23$, $q = 43$, and $d = 485$. A ciphertext message "538" was intercepted. What was the message that was sent?

Standard Decrypting without CRT

The private key $g(x)$ is used to recover the original message.

Since $m = pq = 23 \times 43 = 989$, we have

$$g(x) = g(538) = 538^{485} = 32 \ (\text{mod } 989).$$

The deciphered message is 32.

Decrypting with CRT

Step 1

With the implementation of CRT, $g(x)$ is rewritten as a system of equations. The moduli m ($m = pq = 989$) is replaced with two prime numbers, $p=23$ and $q=43$.

$$g(x) = x^{485} = 538^{485} (\text{mod } 989) \implies \begin{cases} g(x) = 538^{485} (\text{mod } 23) \\ g(x) = 538^{485} (\text{mod } 43) \end{cases}$$

then

$$\begin{cases} g(x) = 538^{485} = 9^{485} = 9 \ (\text{mod } 23) \\ g(x) = 538^{485} = 22^{485} = 32 \ (\text{mod } 43) \end{cases}$$

Step 2

Next, solve the following equations with CRT,

$$\begin{cases} g(x) = 9 \ (\text{mod } 23) \\ g(x) = 32 \ (\text{mod } 43) \end{cases}$$

yields $g(x) = 32 \ (\text{mod } 989)$.

The deciphered message is 32.

The RSA-CRT approach is nearly four times faster than the standard RSA algorithm alone without CRT, (Shinde and Fadewar, 2008 and Stalling, 2010).

Conclusion

The Chinese Remainder Theorem, CRT, is a jewel of mathematics. It can be used to replace a difficult problem with several easy problems. Its usefulness has been demonstrated in the fields of coding, computing and cryptography. Other notable applications can also be found in sequence numbering, Fast Fourier Transform, Dedekind's Theorem, range ambiguity resolution, determination of the greatest common divisor of polynomials and the inverse of the Hilbert matrix. Although known for centuries, CRT continues to present itself as a new application in new contexts and open vistas.

References – Chinese Counting

[1] Gilbert, L. and J. Gilbert. Elements of Modern Algebtra, 8th Edition, Brooks/Cole, Cengage Learning, CA, 2015

[2] Hungerford, T. W. Abstract Algebra: In Introduction. Saunders College Publishing, 1990

[3] Hardy, D. W. and C. L. Walker. Applied Algebra: Codes, Ciphers, and Discrete Algorithms, Pearson Education, Inc., NJ, 2003

[4] Shinde, G. N. and H. S. Fadewar. Faster RSA Algorithm for Decryption Using Chinese Remainder Theorem, vol. 5, no. 4, pp255-261, ICCES, 2008

[5] Stallings, W. Cryptography and Network Security: Principles and Practice, 5th Edition, Prentice-Hall, NJ, 2010

[6] Sun Tsu; The Mathematical Classic of Sun Tsu,孫子算經, China, 400 A.D.

9
The Quest for Symmetry

The Evolution of Scientific Methods

Aristotelian inductive-deductive reasoning was an early scientific way of thinking. Aristotle used inductive reasoning from observations to infer a general statement. Using deductions from general statements, he reached logical conclusions.

Aristotle (384 - 322 BC) stressed the importance of observation in science.

Deductive reasoning, also known as logical deduction, links premises to reach conclusions. If the premises are true, and the rules of deduction are followed, the conclusions must be true.

Dogs like bones.

Archie is a dog.
So Archie must like bones.

But be aware of deductive fallacies:

The King is always right.

The King says kitty is a dog.
Therefore, kitty is a dog.
Long live the King!

Galileo's Experiments and Mathematics

Prior to Galileo's scientific method, Aristotle devised a method of reasoning using deductions from assumptions rather than experimentations.

Galileo constructed his first telescope in 1609, refined after models first devised in the Netherlands. He later made an improved version that could magnify objects up to twenty times. With this telescope, he was able to observe the moon and other heavenly bodies*.

He discovered that the moon was not what was long thought to have been a translucent and perfect sphere, as Aristotle claimed.

© 2017 Chih Yang

Galileo Galilei (1564 - 1642) used experiments and mathematics as research tools. He changed research methodologies from a verbal, qualitative approach to a quantitative one.

Galileo dropped two objects of different weights from the Leaning Tower of Pisa. Through this experiment, he disproved the Aristotelian view that heavy objects fall faster than lighter ones. He also developed a mathematical formula for free-falling objects.

* Warning: Never look directly at the sun through a telescope or binoculars.

Isaac Newton's Giant Step

Isaac Newton (1643-1727) was a key contributor to the scientific revolution in the 16th and 17th centuries.

In the seventeenth century, there were two scientific methods before Newton. One was Francis Bacon's empirical "Inductive Method." The other was Rene Descartes' rational "Deductive Method."

Newton combined these methods and refined the process into one we still use today:

- Conducting experiments
- Collecting and analyzing data
- Formulating a hypothesis
- Testing the hypothesis with additional experiments
- At every stage, describing the procedures so that others can replicate the experiment

The Breakdown of Classical Physics

The success of Newton's natural laws led many to believe that the description of the universe was complete. However, the universe turned out to be more complex than Newtonian mechanics could explain.

By the turn of the 20th century, several large cracks were threatening the foundation of classical physics. Among them were the following:

- James Clerk Maxwell's Theory of electrodynamics
- The Michelson-Morley experiment
- The discovery of X-rays and radioactivity

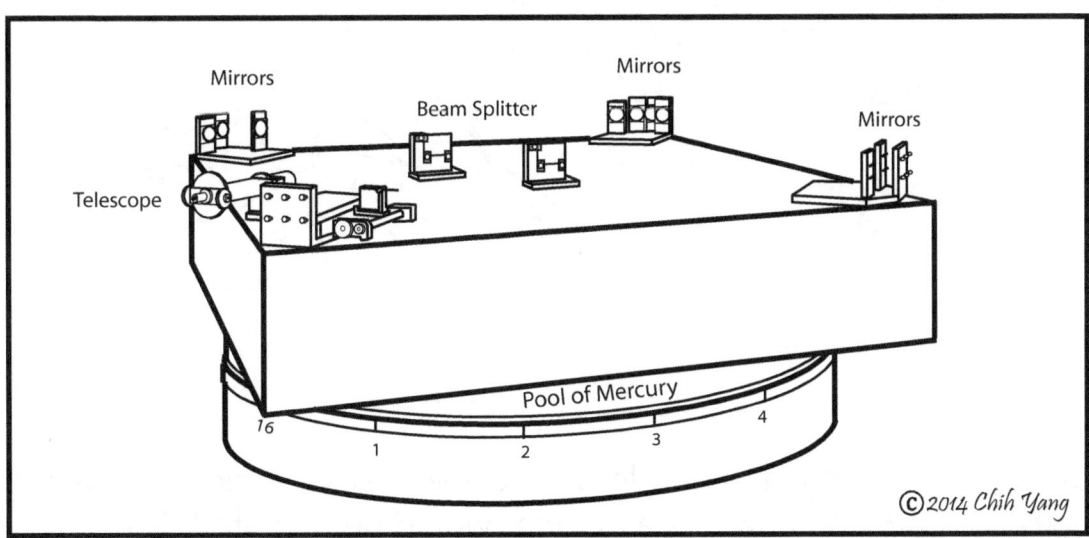

The aim of the Michelson-Morley experiment in 1887 was to uncover the presence and properties of a substance called "luminiferous ether", a substance believed to fill empty space. It was regarded as the most famous failed experiment and was considered to be the first strong evidence against the existence of luminiferous ether. Its null conclusion contradicted Newtonian mechanics.

Later on, scientists recognized that Newtonian mechanics alone was not adequate to describe the fundamental principles of all natural phenomena.

Rocky Road

Post-Galileo, physicists had been guided by the outcomes of experimentation for nearly 400 years. However, by the end of the 19th century, theories in physics were no longer shaped by experiments alone.

Some scientists began to ponder where new ideas for scientific theories would come from and how these theories would be validated. Scientists eagerly began searching for new guiding principles.

Aesthetics

At the dawn of the 20th century, aesthetics emerged as a new guiding principle led by Albert Einstein.

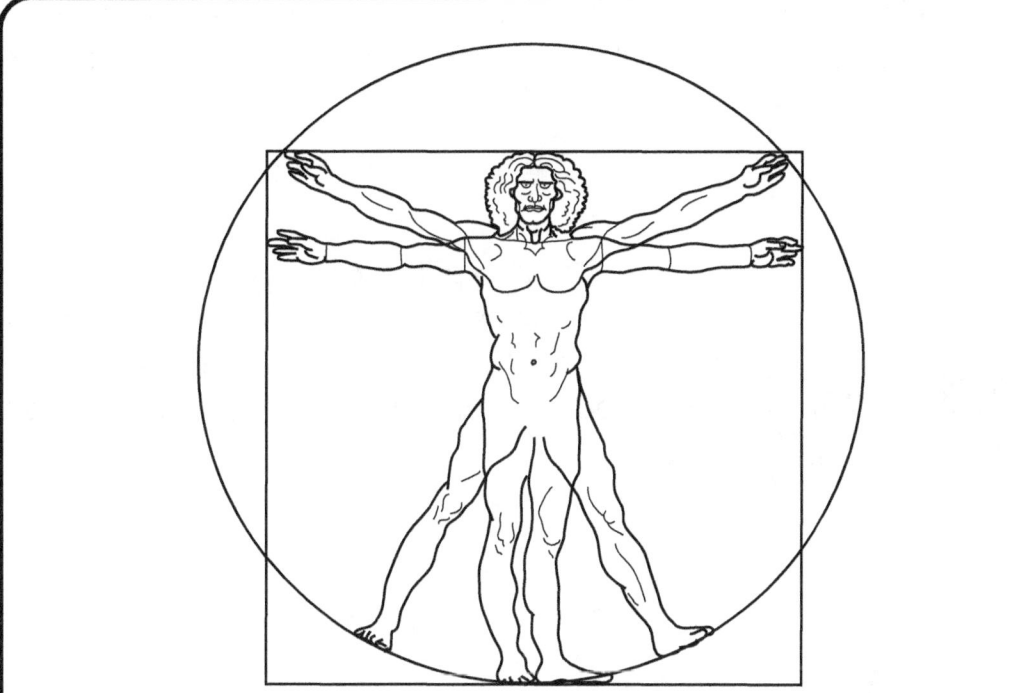

Leonardo da Vinci's "Vitruvian Man" represents the symmetry found in the human body and in the natural universe. A man is fit into a circle and a square by moving his limbs. The area of the circle and the square are the same. They are **invariant**.

Aesthetics is a branch of philosophy focusing on the nature of art and appreciation for beauty. The exploration of aesthetics inevitably led to the discussion of **symmetry**. Symmetry refers to a harmonious balance of parts or elements. According to Herman Weyl, a 20th century mathematician:

"Symmetry, as wide or as narrow as you may define its meaning, is one idea by which man through the ages has tried to comprehend and create order, beauty and perfection."

The Quest for Symmetry

Einstein regarded symmetry as a principle in understanding the natural world. In the 1920s, with the development of quantum mechanics, the symmetry principle became increasingly important in the exploration of modern physics. Today symmetry is a guiding principle in the theoretical formulation of physical laws.

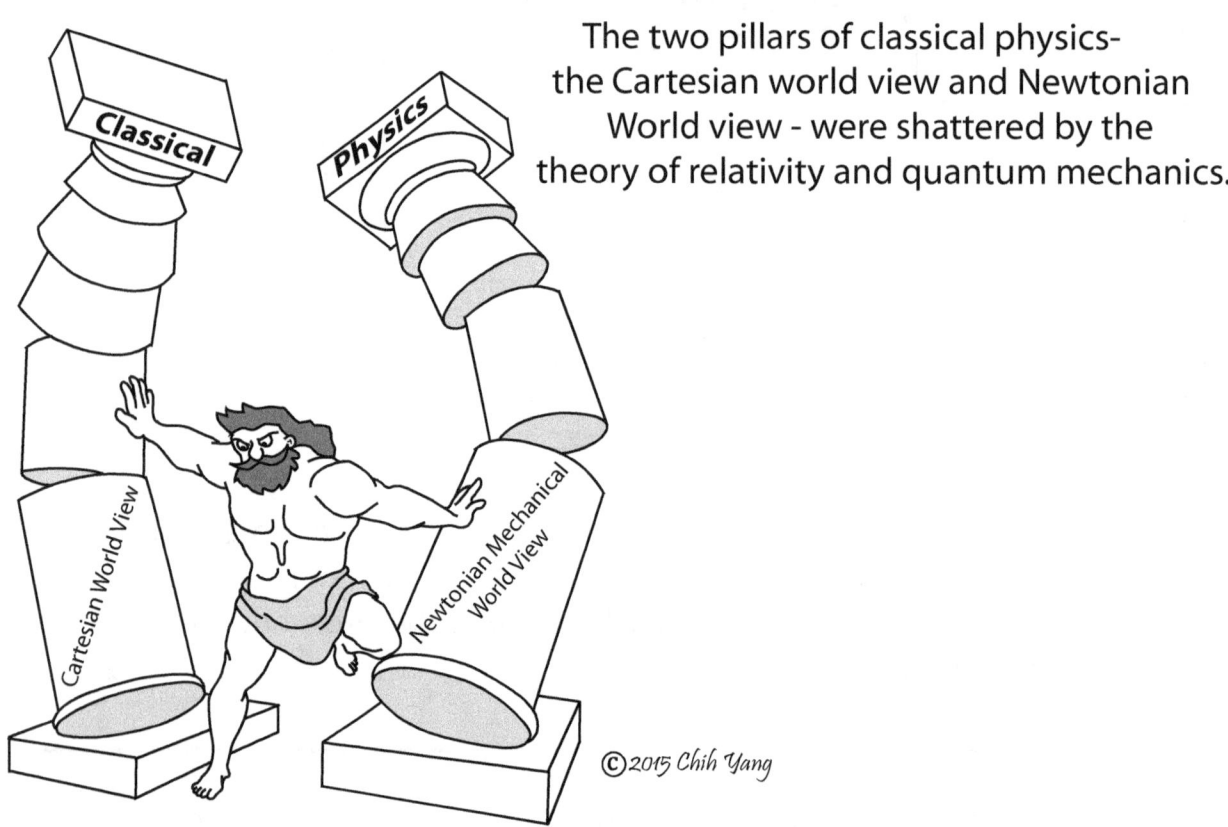

The two pillars of classical physics - the Cartesian world view and Newtonian World view - were shattered by the theory of relativity and quantum mechanics.

The search for new physics theories turned out to be a quest for symmetry.

Symmetry

Symmetry is a one-to-one correspondence of parts or elements on opposite sides of a center, axis or plane. Symmetry is found everywhere: in nature, in design, in art and architecture.

Symmetry in Nature

The concept of symmetry is not strictly limited to geometrical symmetry. There may be other forms of symmetry as well.

Symmetry in Abstract Terms

In mathematics, symmetry is a rigid motion of an object - a transformation, rotation, translation or reflection of an object that leaves the object unchanged, or **invariant**.

In physics, symmetry refers to a property of a physical system that is invariant after a certain transformation.

Groups of Symmetry

In two dimensional symmetry, there are two kinds of groups: frieze groups and non-frieze groups. Freize groups repeat a pattern in one direction. They are **line invariant**.

(1) Frieze Groups - Translations and glide reflections:

There are exactly seven possible types.

(2) Non-frieze Groups - Wallpaper Groups

Non-frieze groups are not line invariant. In non-frieze groups, there are 17 distinct patterns. Such patterns are familiar in decorative art and architecture.

©2015 Chih Yang

(Non-frieze Groups continued)

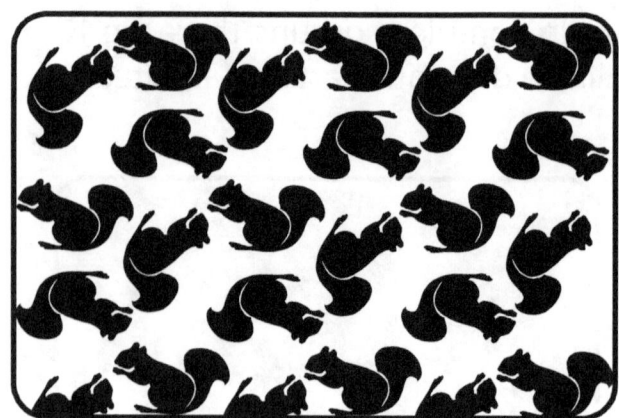

©2015 Chih Yang

(Non-frieze Groups continued)

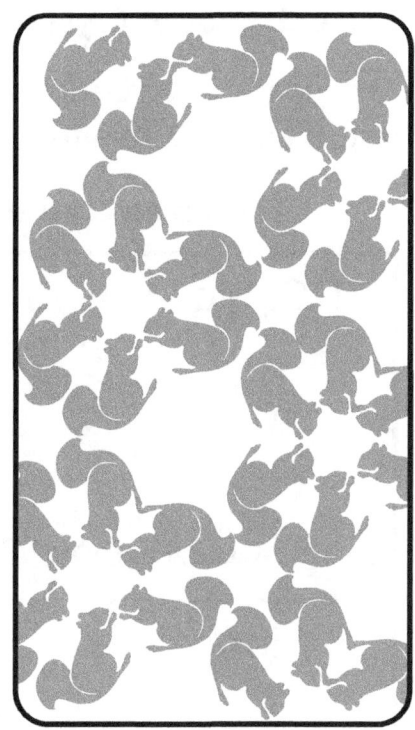
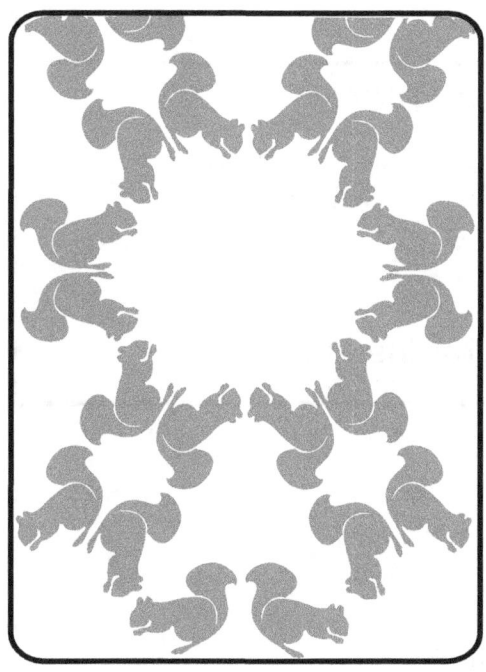
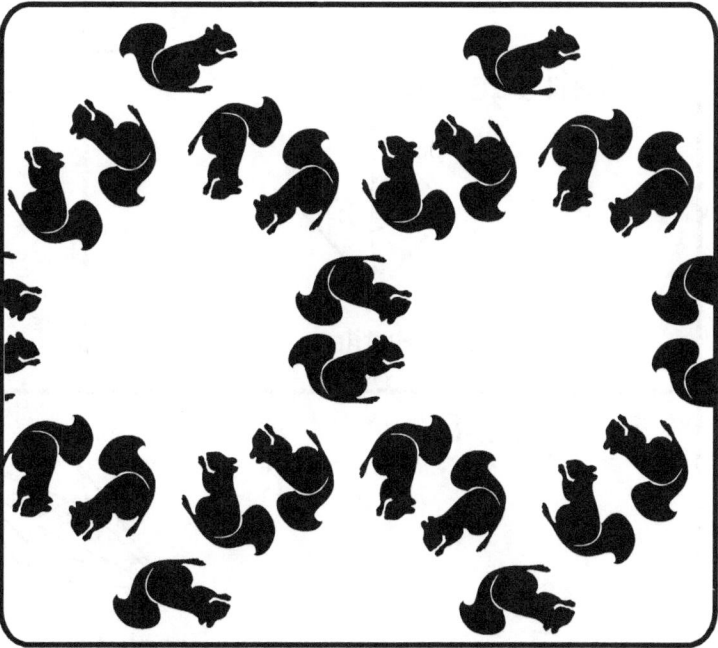

©2015 Chih Yang

Three-Dimensional Symmetry

Space Groups or Crystallographic Groups

There are seven crystal systems which are classified into 32 finite symmetric groups and 230 infinite symmetric groups.

The seven crystal systems with examples of geometric shapes:

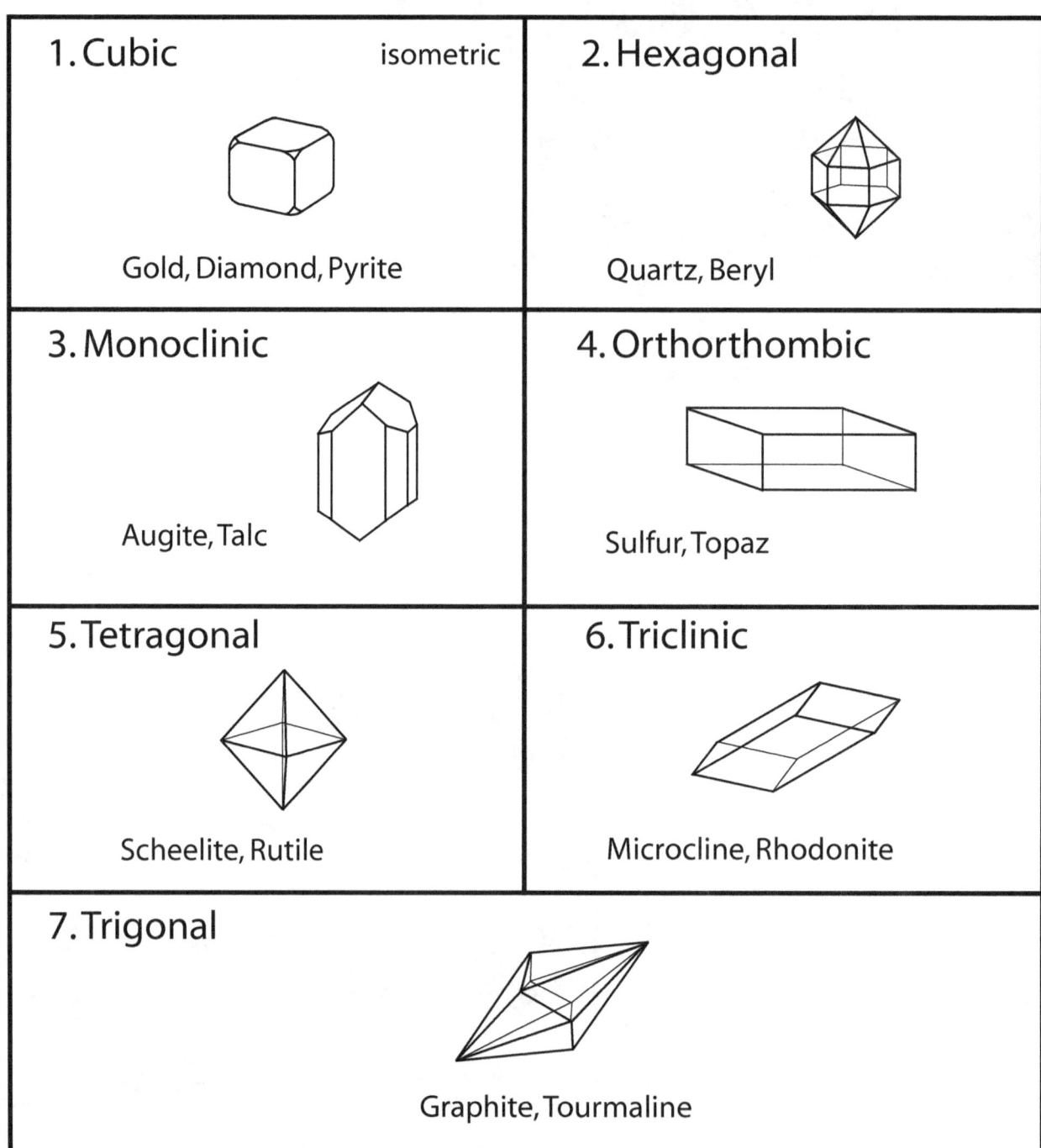

Chirality

An object is said to be "chiral" if it cannot be superimposed on its mirror image. Chirality is a property of asymmetry. It shows handedness.

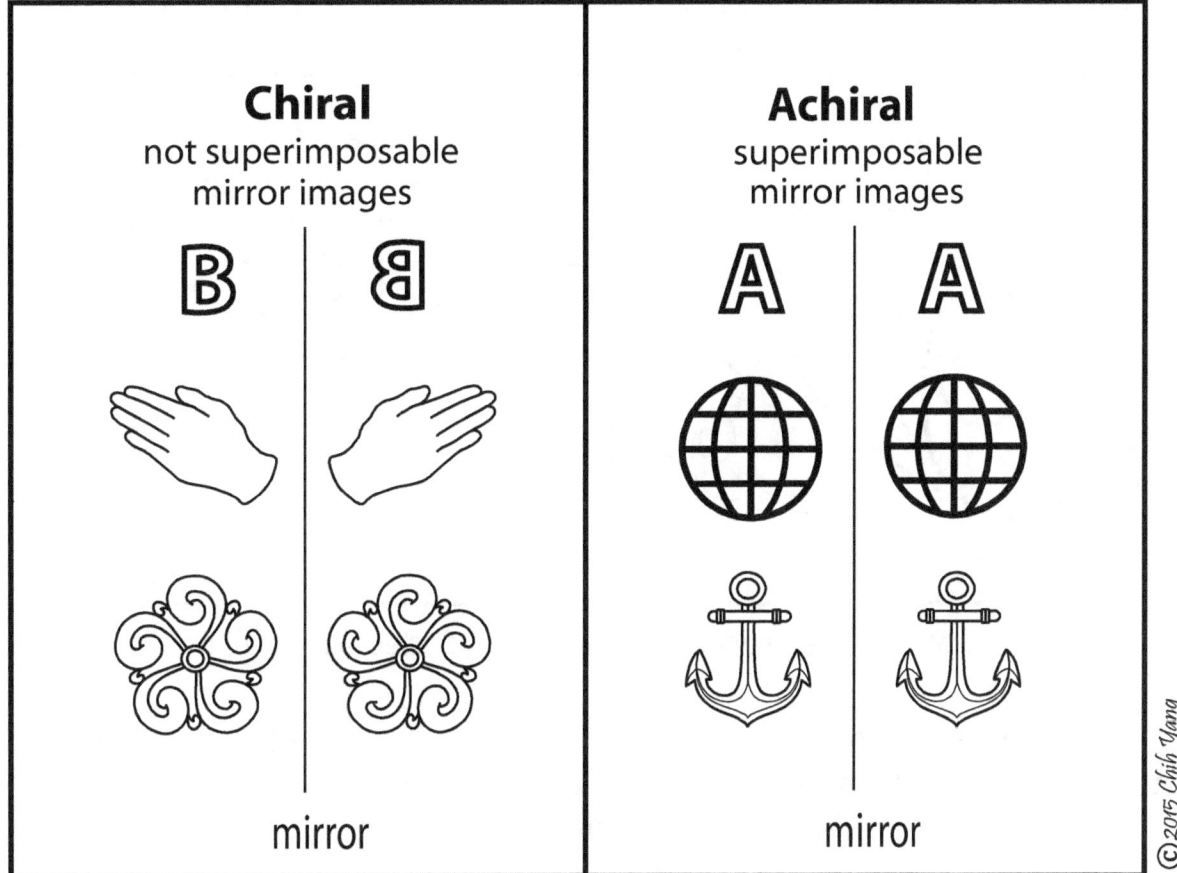

In 1848, Louis Pasteur discovered the existence of chirality in experiments with crystallization. Pasteur found two kinds of tiny crystals of the same substance, crystals that were mirror images of each other but not superimposable. One type of crystal was found to be identical to the tartaric acid present in fermenting grapes. The other type of crystal had never been observed before in nature.

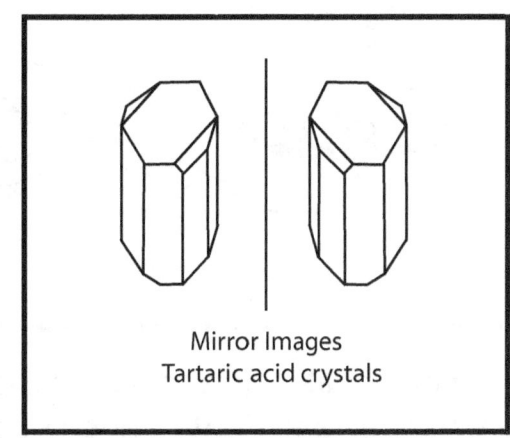

Mirror Images
Tartaric acid crystals

Chirality in Nature

Nearly all biological polymers must have the same handedness to function and support life. All amino acids in living organisms occur as left-handed. On the other hand, all nucleotides in RNA and DNA are right-handed.

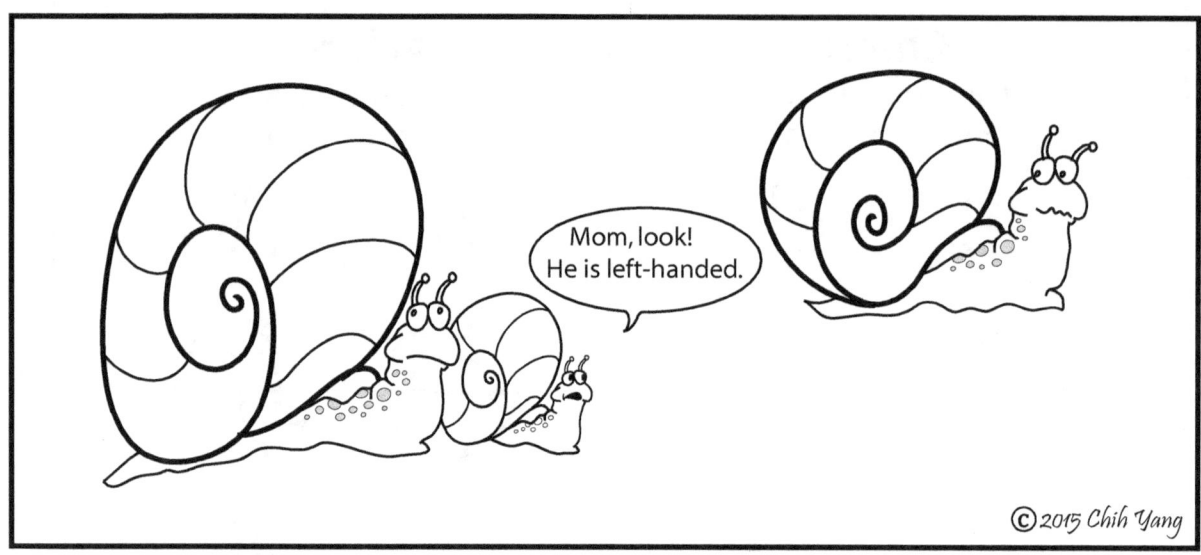

The majority of gastropods have right-handed (dextral) shells.

Homochirality - In nature, some substances are always left-handed and others always right-handed. A substance is said to be homochiral if all its constituent parts exhibit the same chiral form.

Chirality in Ordinary Chemical Synthesis

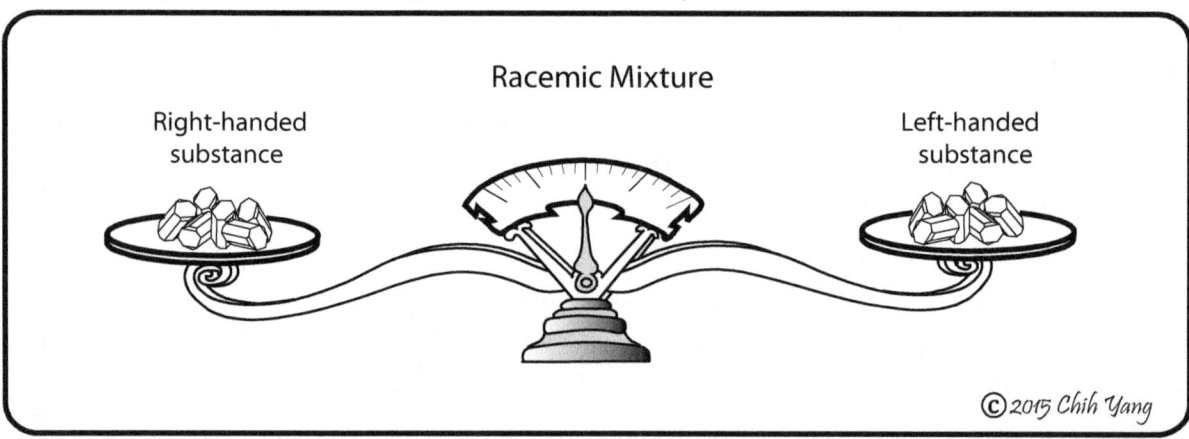

When chiral substances are produced in the laboratory, both versions - left-handed and right-handed - occur in equal amounts (a 50/50 mixture, called a racemic mixture).

A Tragic Example of the Importance of Chirality: Thalidomide

Thalidomide was a drug introduced in the 1950s in Western Europe under the name of Contergan. Physicians prescribed it as a sedative for pregnant women. This drug was made and sold as a racemic mixture of R-thalidomide and S-thalidomide.

Mirror Images of Thalidomide

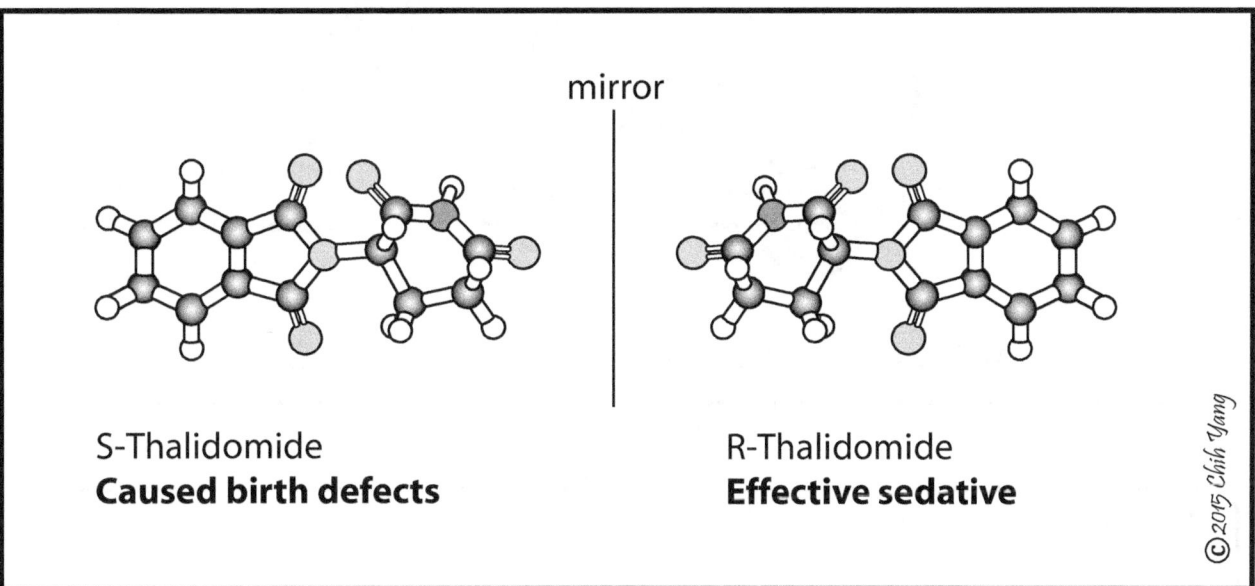

Thalidomide -- which resulted in thousands of babies born with missing or distorted arms, legs, hands or feet -- was banned in 1961.

A birth defect caused by thalidomide

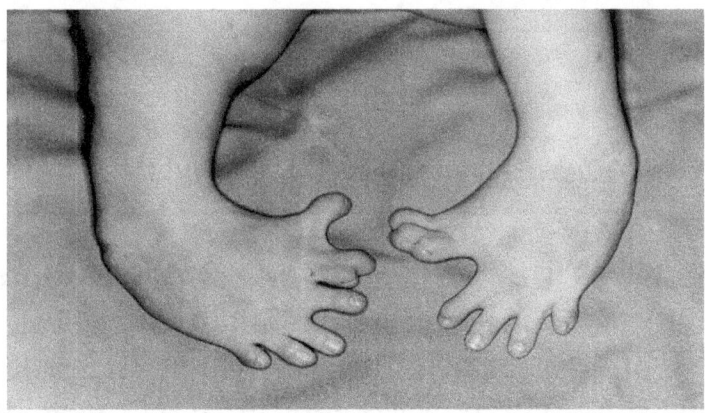

(Source: http://commons.wikimedia.org/wiki/File:NCP14053.jpg)

Symmetry in Physics

Conservation of Momentum

In a closed system, total momentum is constant. It does not change.

Before

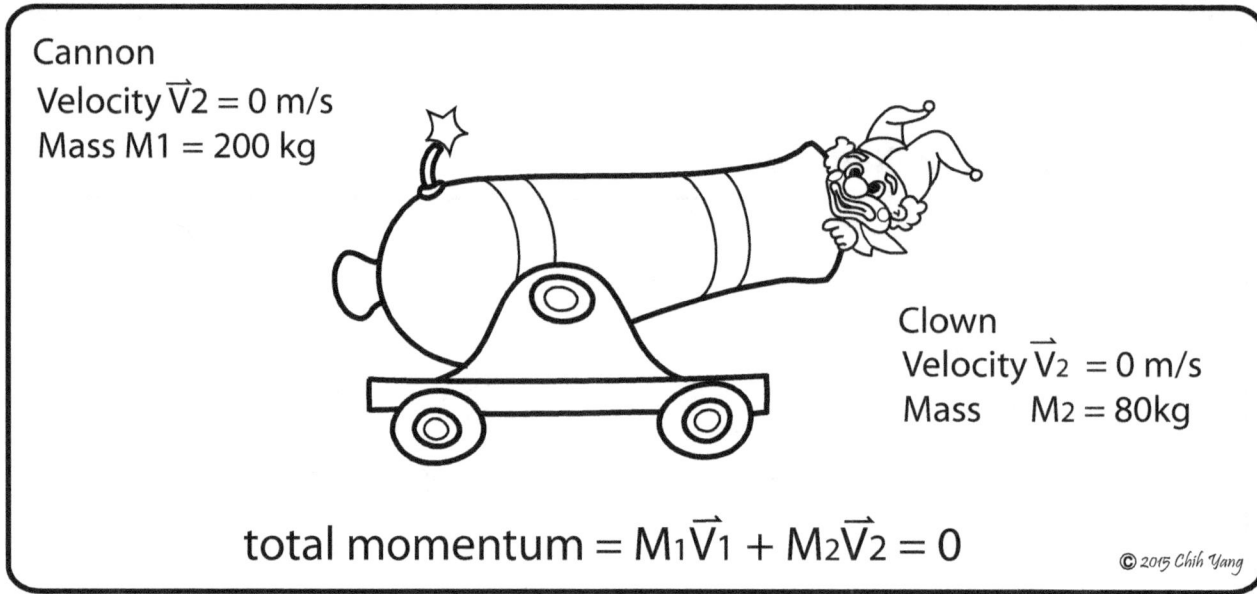

Cannon
Velocity $\vec{V}_2 = 0$ m/s
Mass $M_1 = 200$ kg

Clown
Velocity $\vec{V}_2 = 0$ m/s
Mass $M_2 = 80$ kg

total momentum = $M_1\vec{V}_1 + M_2\vec{V}_2 = 0$

© 2015 Chih Yang

After

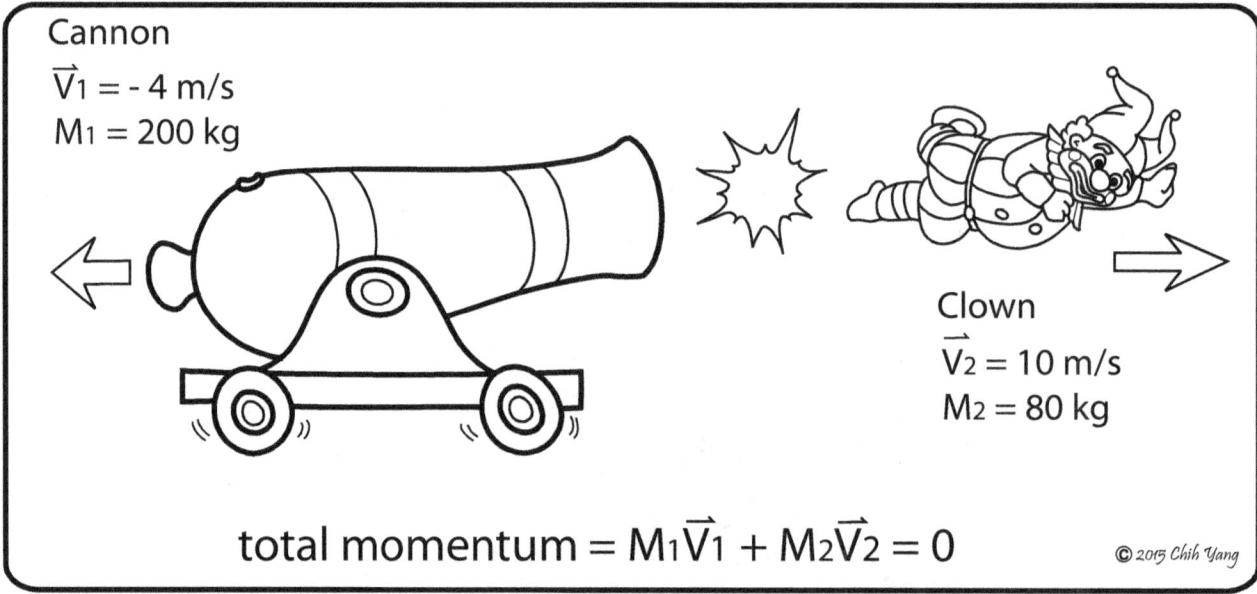

Cannon
$\vec{V}_1 = -4$ m/s
$M_1 = 200$ kg

Clown
$\vec{V}_2 = 10$ m/s
$M_2 = 80$ kg

total momentum = $M_1\vec{V}_1 + M_2\vec{V}_2 = 0$

© 2015 Chih Yang

According to Noether's Theorem, in every differentiable **symmetry** generated by local actions, there are corresponding quantities whose values are conserved throughout the time of the action. This theorem in physics states the Laws of Conservation of Energy, Conservation of Momentum and Conservation of Angular Momentum.

Symmetry in the Sky

The Big Dipper

The Big Dipper is the best known asterism in the northern skies. The Big Dipper demonstrates a rotational symmetry as it appears to move counter-clockwise around the celestial North Pole.

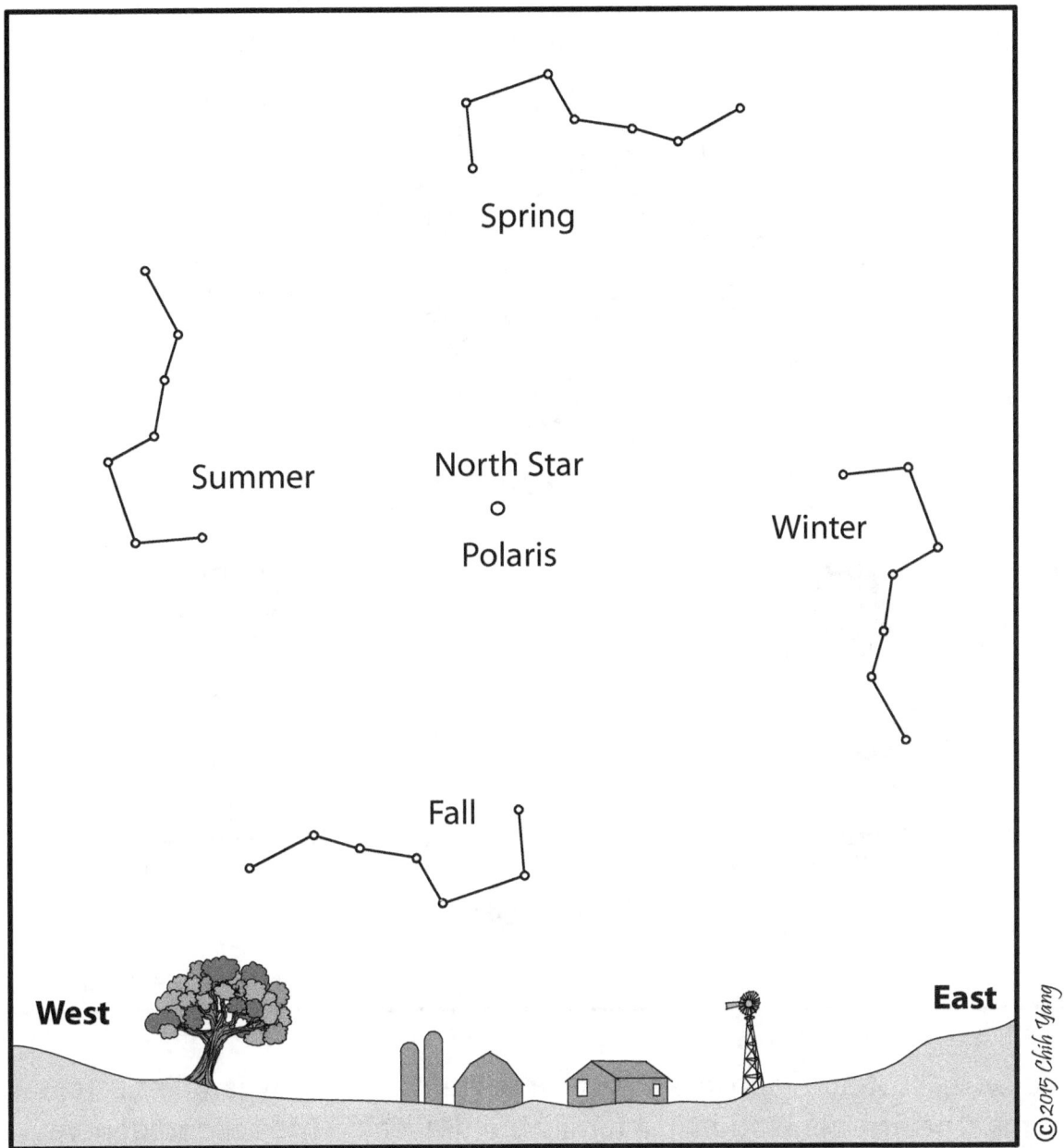

The early evening position of the Big Dipper relative to the position of the North Star at different seasons of the year

Because of the rotation of the earth, the Big Dipper appears to circle the North Star once every 24 hours.

Symmetry in Wealth Management

From Rags to Rags in Three Generations

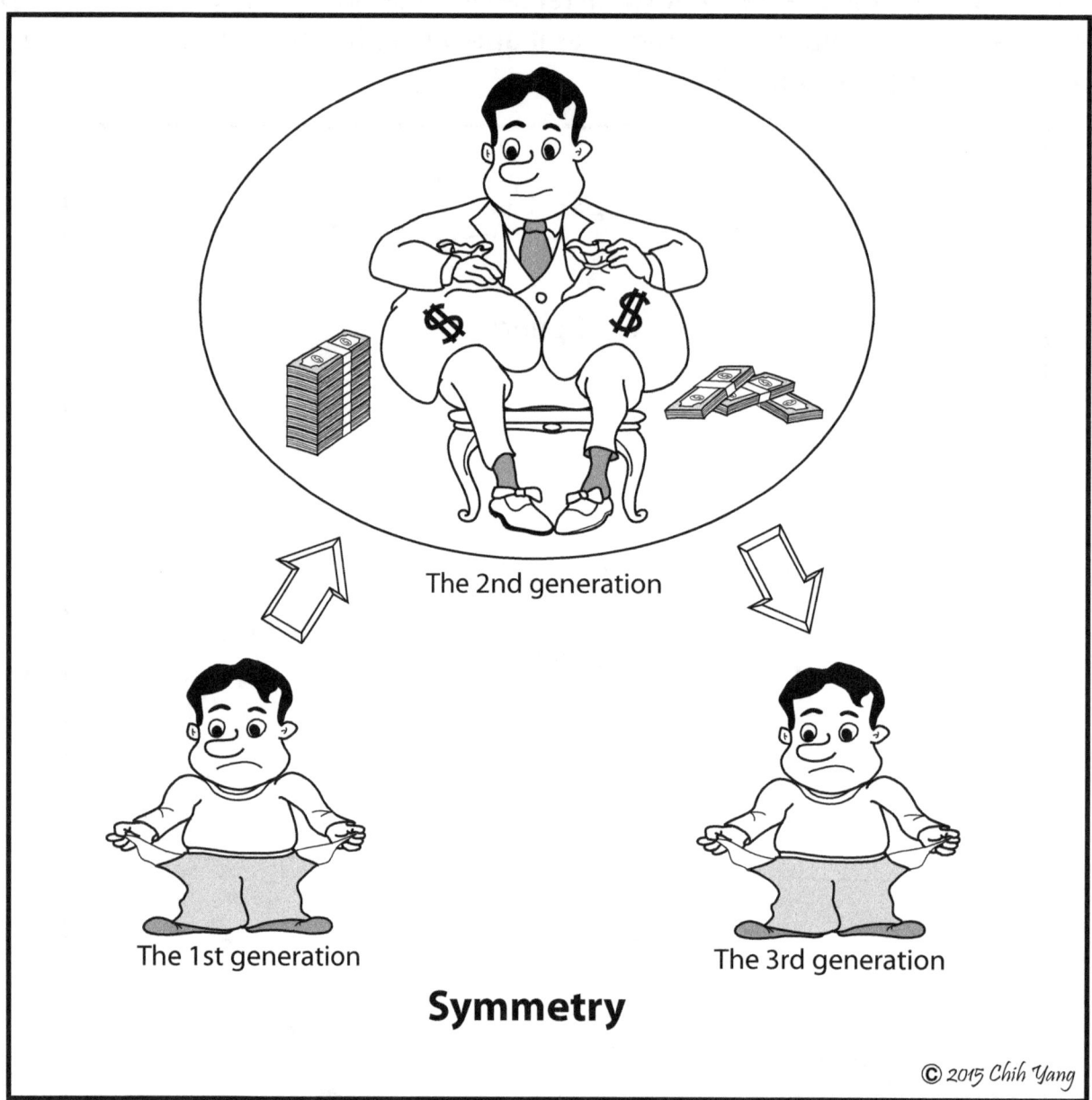

The cycle illustrated above is a familiar pattern in wealthy families across many cultures. The first generation in a family builds wealth. The second generation shepherds or preserves it. The third generation squanders it. According to statistics, 90% of wealthy families have lost their fortune by the end of the third generation.

In the financial services industry, some services are designed to help the wealthy break this cycle.

Symmetry in Music

Music is full of examples of symmetry. Among the most familiar are Johann Sebastian Bach's canons and fugues. The canon and fugue are contrapuntal compositional techniques. The canon introduces a melody which, after a given interval, is followed by one or more imitations of the melody. These imitations may be played on different instruments or sung in a different vocal range.

Musical Quotation from Pachelbel's Canon in D

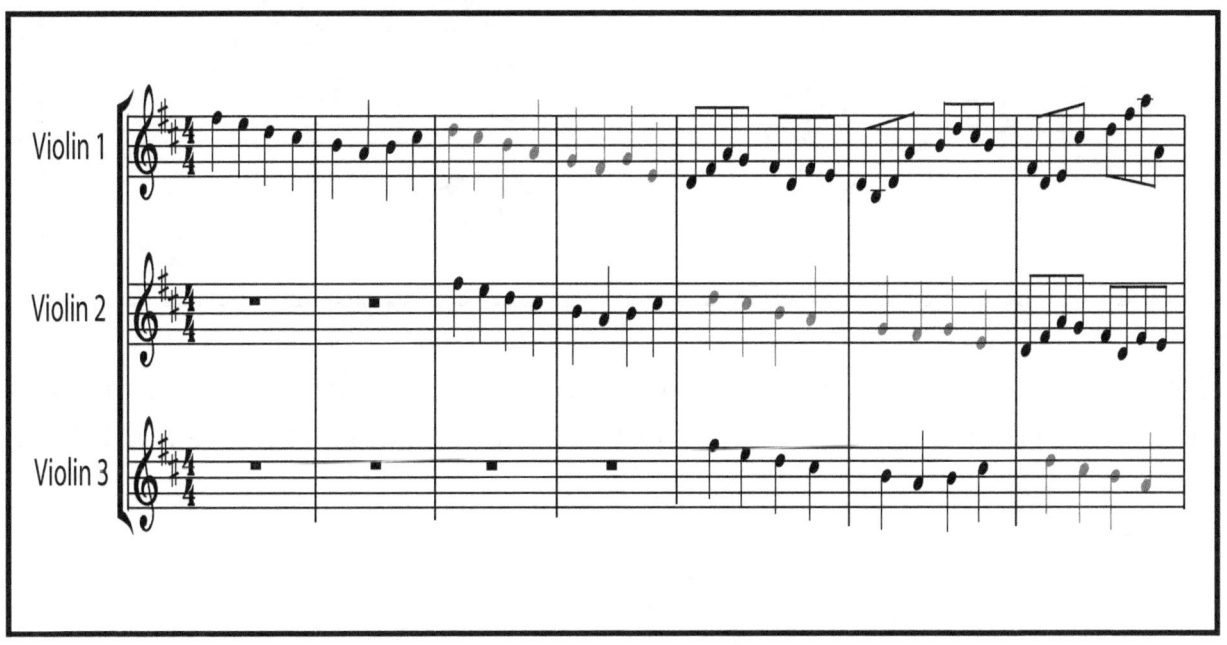

Symmetry in Image Recognition

Example: Traffic Information

It's a typical morning rush hour. The traffic control system is busy as usual.

How can the computer recognize that the data associated with a car that was in the left lane just moments ago is now associated with a car in the right lane?

The traffic pattern is a symmetry with respect to time. For a computer to recognize and interpret the images, the Group Theory presents a mathematical solution. **Group Theory** is the language of symmetry.

References – Symmetry

[1] Booth, Basil; Rocks and Minerals; Chartwell Books, Inc., NJ, 1993

[2] Durbin, John R.; Modern Algebra-An Introduction, 5th Edition, John Wiley & Sons, Inc., 2005

[3] Farris, Frank A.; Creating Symmetry: The Artful Mathematics of Wallpaper Patterns, Princeton University Press, NJ, 2015

[4] Gross, David J.; The Role of Symmetry in Fundamental Physics, Proceedings of the National Academy of Sciences of the United States of America, 1996

[5] Hargittai, Magdoina and Istvan Hargittai; Visual Symmetry, World Scientific Publishing, NJ, 2009

[6] Lemon, Harvey Bruce; From Galileo to the Nuclear Age, 2nd Edition, The University of Chicago Press, Chicago, 1946

[7] Polya, George; Mathematical Methods in Science, Mathematical Association of America, Washington D.C., 2012

[8] Ronan, Mark; Symmetry and the Monster: One of the Greatest Quest of Mathematics, Oxford University Press, Oxford, 2006

[9] Selig, J.M.; Geometric Fundamentals of Robotics, 2nd Edition, Springer, 2005

[10] Tipler, Paul A. and Ralph A. Llewellyn; Modern Physics, 6th Edition, W.H. Freeman and Company, NY, 2012

[11] Weyl, Hermann; Symmetry, Princeton University Press, 1952

[12] Webb, Stephen; Out of This World: Colliding Universe, Branes, Strings, and Other Wild Ideas of Modern Physics, 2004, Copernicus Books, 2004

[13] Zee, A. and Roger Penrose; Fearful Symmetry: The Search for Beauty in Modern Physics, Princeton Science Library, NY, 2007

10

Algebraic Systems

Algebraic Systems

An *algebraic system* is a set with one or more operations defined on it. Those operations must conform to algebraic laws.

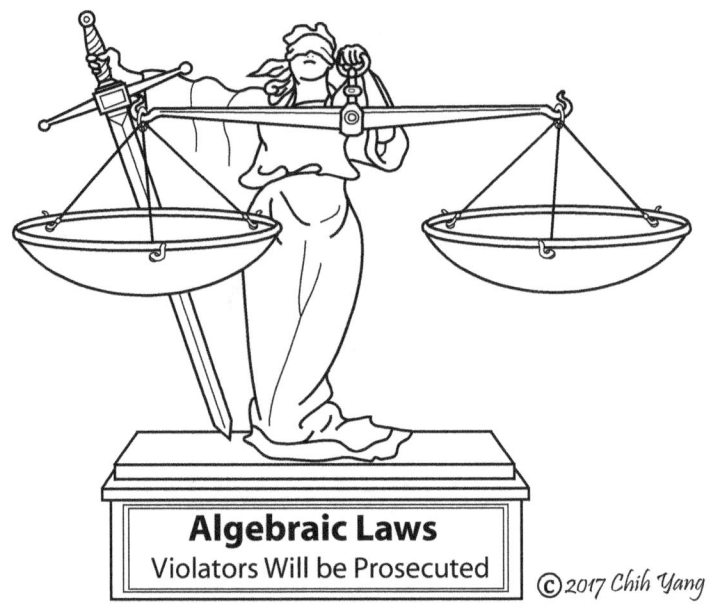

Algebraic Laws

Suppose that **A** is a set in which two binary operations are defined. We will write the operations as addition and multiplication, denoted by + and · respectively. "Addition" does not necessarily refer to the conventional addition of numbers, nor does multiplication necessarily refer to the conventional operation of multiplying numbers.

A1 **Closure law of addition:**
Whenever a and b belong to **A**, then $a + b$ is also in **A**.

A2 **Associative law of addition:**
$a + (b + c) = (a + b) + c$ for all a, b, c in **A**.

A3 **Additive identity:**
There exists an element 0 in **A** such that $0 + a = a + 0 = a$ for all a in **A**.

A4 **Additive inverse:**
For each a in **A**, there exists an element $-a$ in **A** such that $a + (-a) = (-a) + a = 0$.

(Algebraic Laws continued)

A5 ***Commutative law of addition:***
$a + b = b + a$ for all a, b in **A**.

M1 ***Closure law of multiplication:***
For all a, b in **A**, $a \cdot b$ is still in **A**.

M2 ***Associative law of multiplication:***
$a \cdot (b \cdot c) = (a \cdot b) \cdot c$ for all a, b, c in **A**.

M3 ***Multiplicative identity:***
There exists an element 1 in **A** such that $1 \cdot a = a \cdot 1$ for all a in **A**.

M4 ***Distributive law:***
$a \cdot (b + c) = a \cdot b + a \cdot c$ and $(a + b) \cdot c = a \cdot c + b \cdot c$ for all a, b, c in **A**.

M5 ***Commutative law of multiplication:***
$a \cdot b = b \cdot a$ for all a, b in **A**.

M6 ***No zero-divisors:***
If a and b belong to **A** and $a \cdot b = 0$, then either $a = 0$ or $b = 0$.

M7 ***Multiplicative inverse:***
For any a in **A** and $a \neq 0$, there exists an element x in A such that $a \cdot x = x \cdot a = 1$.

A commutative law violator

Group, Ring, Integral Domain and Field

Algebraic Laws and Systems

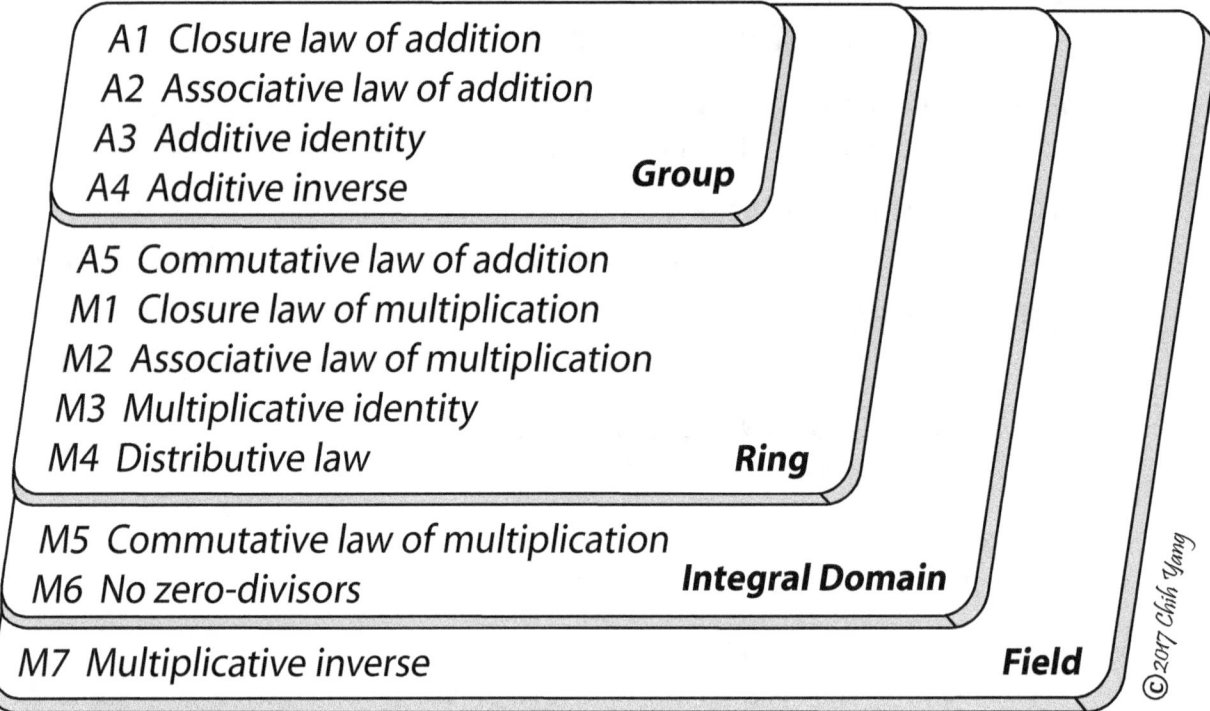

Suppose that **A** is a set in which two binary operations are defined.

(1) If **A** satisfies properties A1 - A4, then **A** forms a **Group** with respect to addition. **A** is called a commutative group or an abelian group, if **A** also satisfies A5.
(2) If **A** satisfies properties A1 - A5 and M1 - M4, then **A** is called a **Ring** under the two given operations. If M5 holds well, **A** is called a commutative ring.
(3) If **A** satisfies properties A1 - A5 and M1 - M6, then **A** is an **Integral Domain**.
(4) If **A** satisfies all properties, A1 - A5 and M1 - M7, then **A** is a **Field**.

Examples
(1) The rational numbers, the real numbers, and the complex numbers, satisfy all properties, are fields.
(2) The set of all integers forms an integral domain.
(3) The set of all even integers forms a commutative ring.
(4) The set of modular numbers Z_{10} is a ring. Since [2] x [5] = [0], [2] and [5] are zero divisors. Z_{10} does not form an integral domain.

Group

Group is a prototype algebraic system consisting of a set with one binary operation that satisfies algebraic laws A1 - A4.

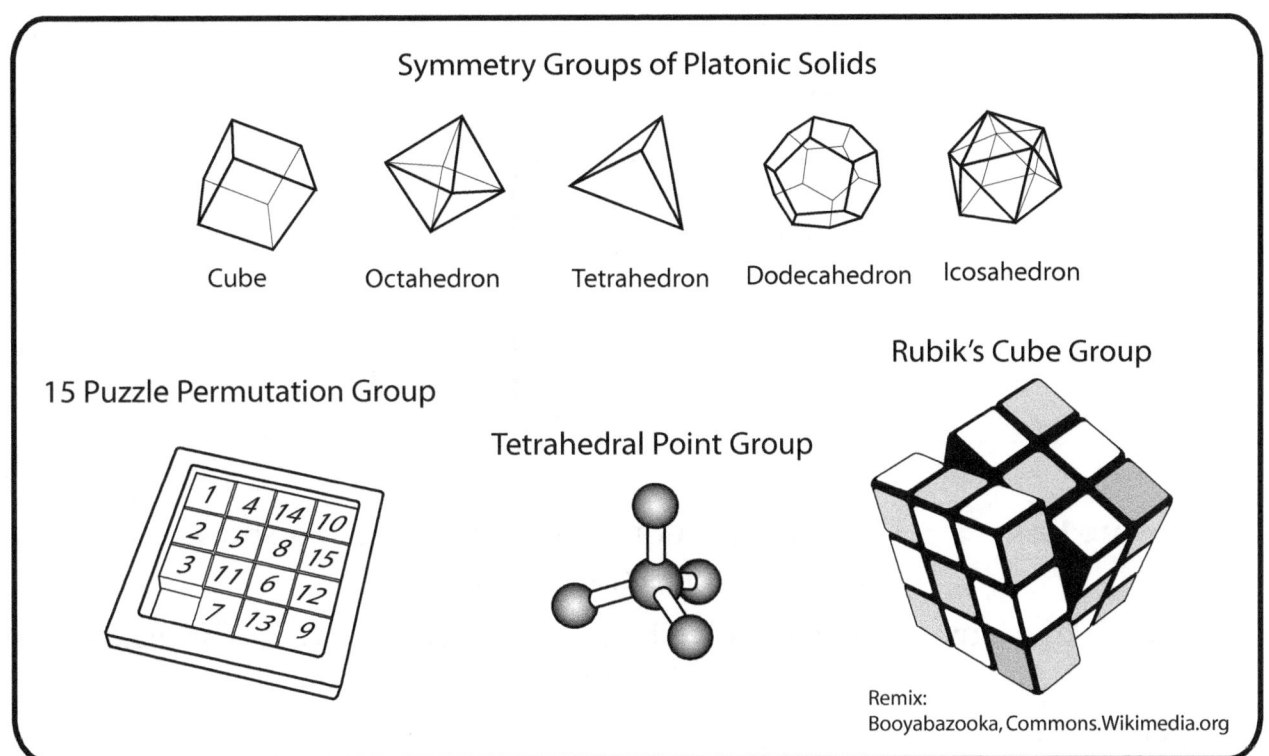

Example 1

The set of integers **Z** forms a group with respect to the operation of conventional addition.

To be a group, the operation on the set must satisfy the following laws:

(1) Closure Law

If a and b are any two integers, the sum $a + b$ must be also an integer.

(2) Associative Law

For any three integers a, b and c, $(a + b) + c = a + (b + c)$.

(3) Identity Law

Zero, 0, is the identity. For any integer a, $a + 0 = 0 + a = a$.

(4) Inverse Law

For each integer a, there exists an inverse integer $-a$ that satisfies $a + (-a) = (-a) + a = 0$.

Example 2

The subset **G**={*1,-1, i, -i*} of complex numbers is a group with respect to the operation of multiplication.

Cayley Table

X	1	-1	i	-i
1	1	-1	i	-i
-1	-1	1	i	-i
i	i	-i	-1	1
-i	-i	i	1	-1

$i = \sqrt{-1}$

(1) Closure Law — The multiplication table, or Cayley table, shows that **G** is closed under multiplication.

(2) Associative Law — **G** is associative because the associative law applies to the set of all complex numbers.

(3) Identity Law — Identity $e = 1$

(4) Inverse Law — Each of *1* and *-1* is its own inverse. *i* and *-i* are inverse of each other.

The Graphic Representation of Group **G**

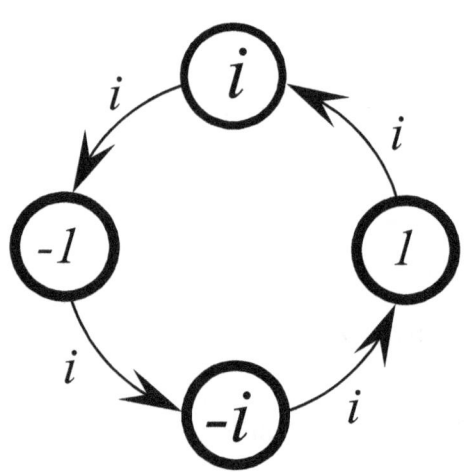

$\xrightarrow{i}$

(A) The arrow means "multiply by *i*."

(B) The element *i* is called the **generator** of **G** since it generates all elements of **G**.

$$\langle i \rangle = \{i^1, i^2, i^3, i^4\} = \{i, -1, -i, 1\} = \mathbf{G}$$

The set of all elements of the form i^n, where n is an integer.

(C) The group $< i >$ is called a **cyclic** group.

Example 3

The integers modulo 5, $\mathbf{Z}_5 = \{[0], [1], [2], [3], [4]\}$, is a group under addition.

Cayley Table

+	[0]	[1]	[2]	[3]	[4]
[0]	[0]	[1]	[2]	[3]	[4]
[1]	[1]	[2]	[3]	[4]	[0]
[2]	[2]	[3]	[4]	[0]	[1]
[3]	[3]	[4]	[0]	[1]	[2]
[4]	[4]	[0]	[1]	[2]	[3]

(1) It is closed.
(2) It is associative
(3) Identity $e = [0]$
(4) Inverse
- [1] and [4] are inverse of each other.
- [2] and [3] are inverse of each other.

The Graphic Representation of Cyclic Group $\mathbf{Z}_5$

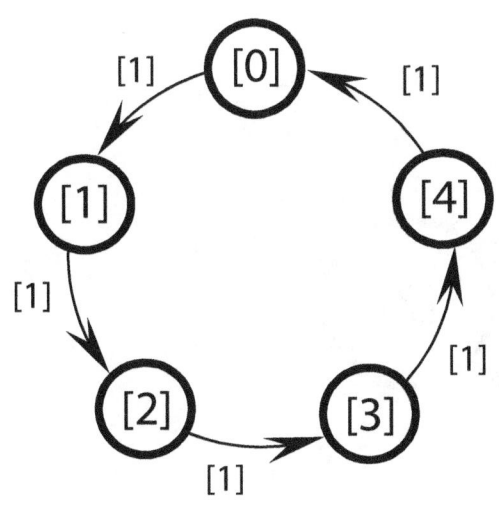

(A) The arrow means "adding [1]."

(B) There are five elements in group $\mathbf{Z}_5$. $\mathbf{Z}_5$ has the **order** of 5.

(C) [1] and [4] are generators of $\mathbf{Z}_5$.

$\langle [1] \rangle = \{ [1]\cdot 1, [1]\cdot 2, [1]\cdot 3, [1]\cdot 4, [1]\cdot 5\} = \mathbf{Z}_5$
$\langle [4] \rangle = \{ [4]\cdot 1, [4]\cdot 2, [4]\cdot 3, [4]\cdot 4, [4]\cdot 5\} = \mathbf{Z}_5$

Example 4

The subset **U**₈ = {[1], [3], [7], [9]} ⊆ **Z**₈, modulo 8, is a group under multiplication

Cayley Table

x	[1]	[3]	[5]	[7]
[1]	[1]	[3]	[5]	[7]
[3]	[3]	[1]	[7]	[5]
[5]	[5]	[7]	[1]	[3]
[7]	[7]	[5]	[3]	[1]

(1) It is closed.
(2) It is associative.
(3) Identity e = [1]
(4) Inverse - Each element is its own inverse.

The Graphic Representation of Group **U**₈

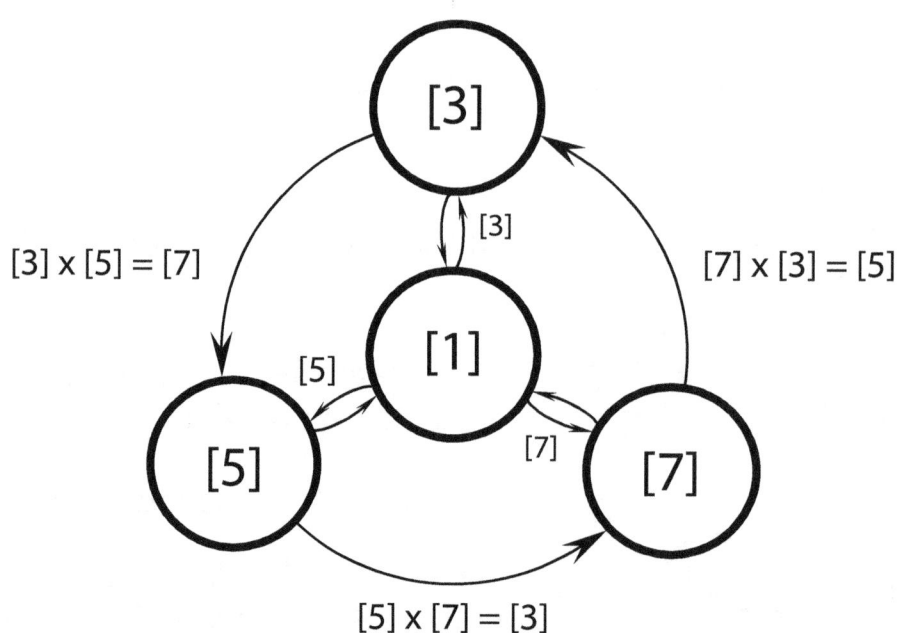

This is called a **Klein Four** group.

No generator.
It's not a cyclic group.

Example 5 Two-way Switching Circuit

In electric wiring, a two-way switching allows you to control a light source from different locations.

There are four possible settings of the switches.

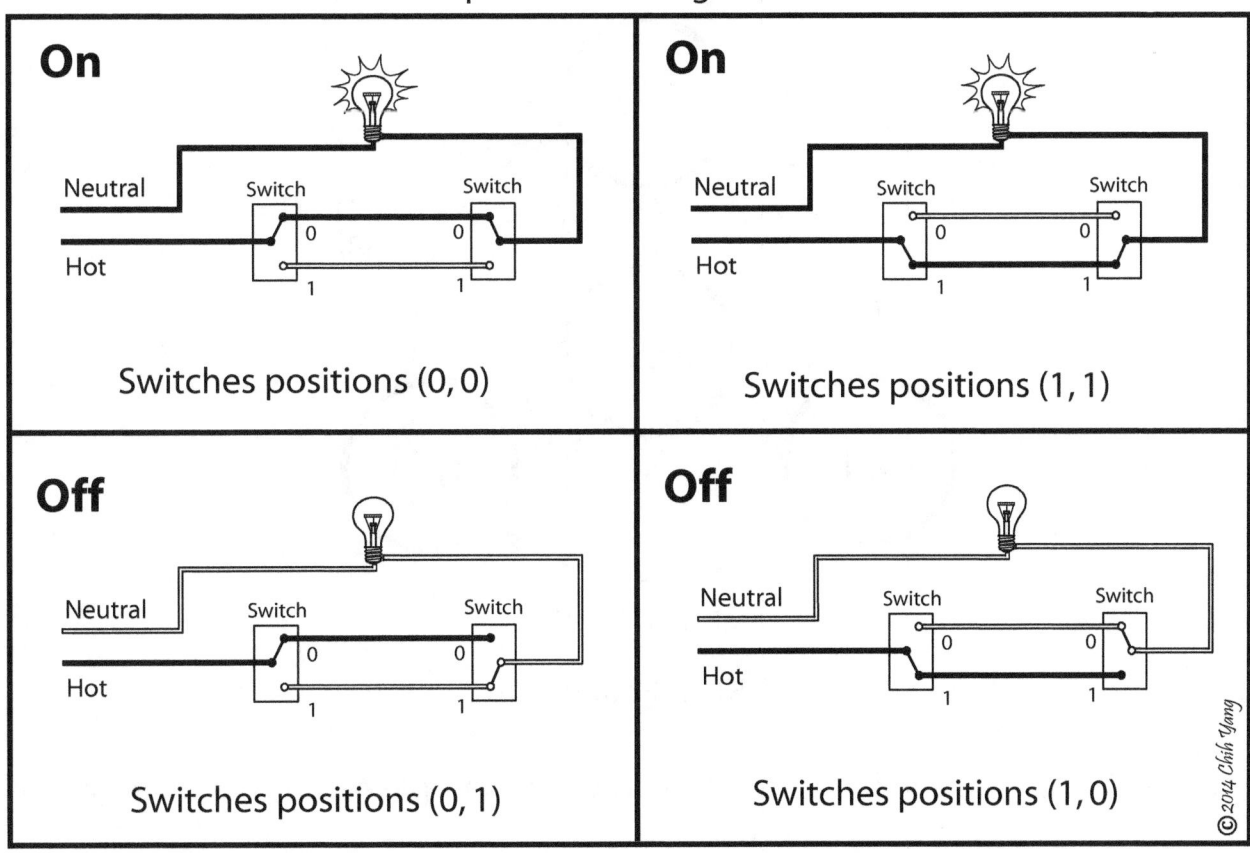

These four positions form a group, $\mathbf{Z}_2 \times \mathbf{Z}_2 = \{(0, 0), (0, 1), (1, 0), (1, 1)\}$.

(Example 5 continued)

$Z_2 \times Z_2 = \{(0,0), (0,1), (1,0), (1,1)\}$ is a group with respect to addition modulo 2.

Cayley Table

+	(0,0)	(0,1)	(1,0)	(1,1)
(0,0)	(0,0)	(0,1)	(1,0)	(1,1)
(0,1)	(0,1)	(0,0)	(1,1)	(1,0)
(1,0)	(1,0)	(1,0)	(0,0)	(0,1)
(1,1)	(1,1)	(1,0)	(0,1)	(0,0)

(1) It is closed.
(2) It is associative.
(3) Identity, $e = (0,0)$
(4) Inverse - Each element is its own inverse.

The Graphic Representation of Group $Z_2 \times Z_2$

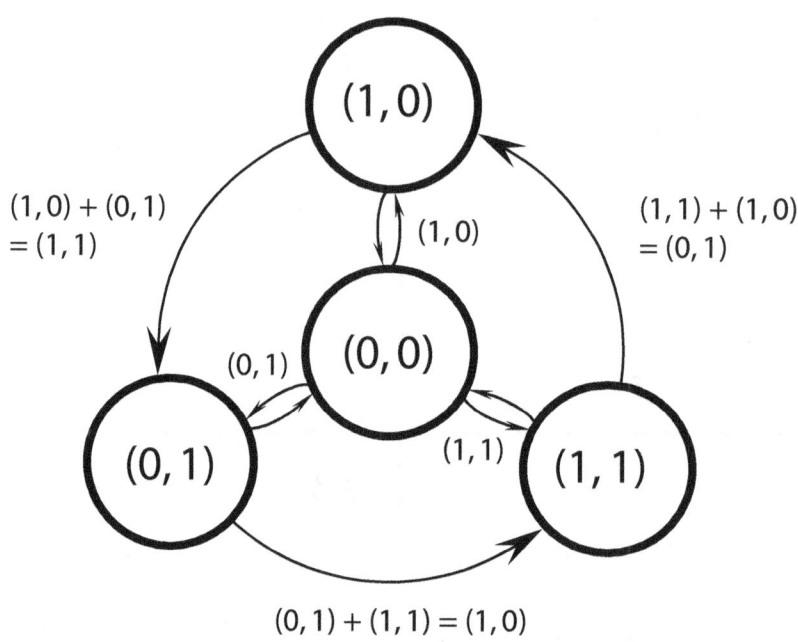

A Klein Four group.

Group $Z_2 \times Z_2$ has the same structure as U_8 in example 4. They are **isomorphic**.

Example 6 The Group of Drill Commands

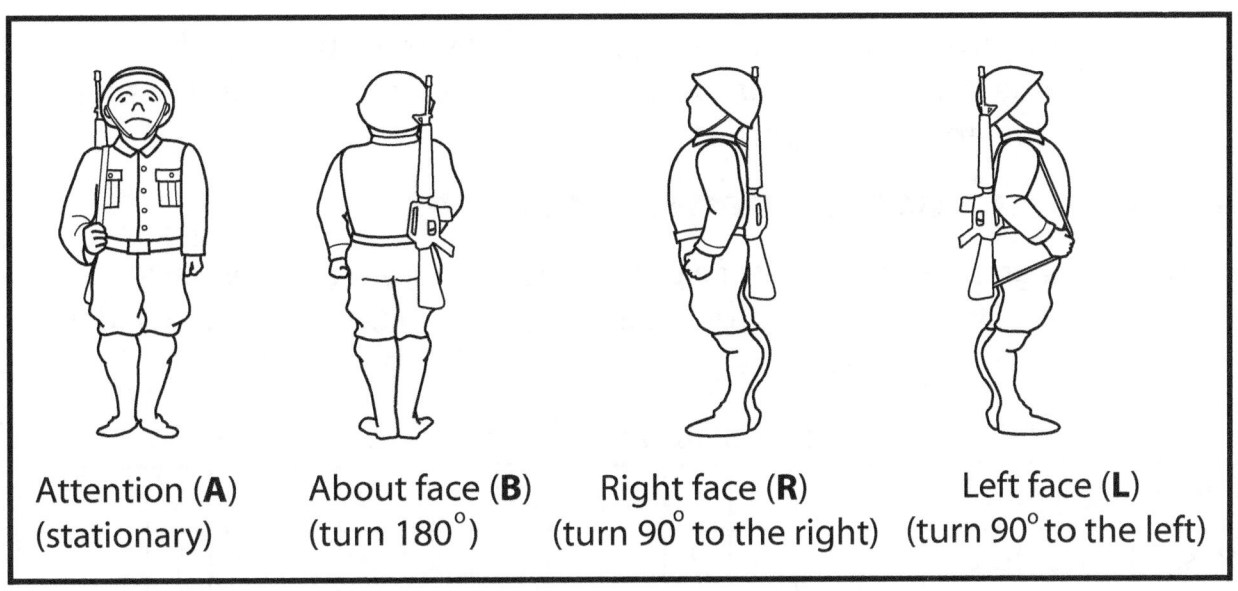

| | Attention (**A**) (stationary) | About face (**B**) (turn 180°) | Right face (**R**) (turn 90° to the right) | Left face (**L**) (turn 90° to the left) |

The above four commands form a set **G**= { **A**, **B**, **R**, **L** }. The operation ⊕ is the action "followed by." Then, **G** is a cyclic group under ⊕.

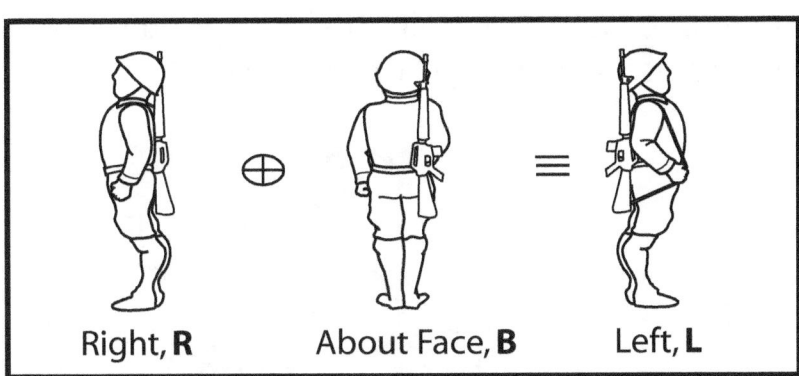

Right, **R**   About Face, **B**   Left, **L**

Cayley Table

⊕	A	B	R	L
A	A	B	R	L
B	B	A	L	R
R	R	L	B	A
L	L	R	A	B

Cayley table shows that **A** is the identity.

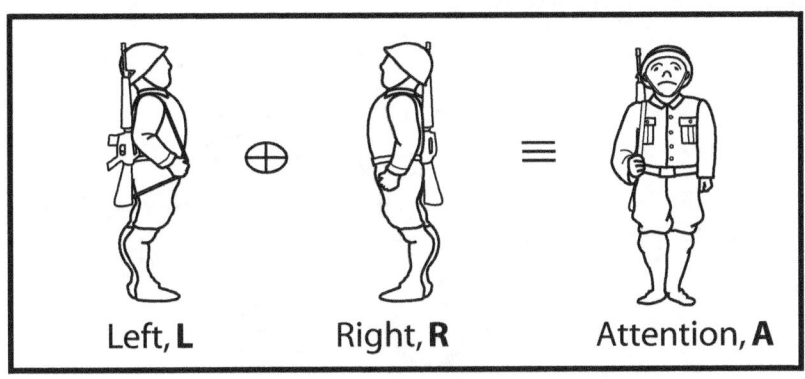

Left, **L**   Right, **R**   Attention, **A**

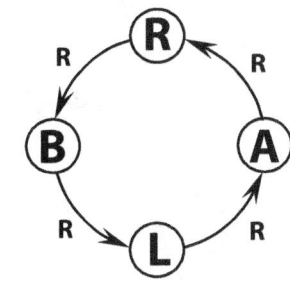

R and **L** are generators.

Classification of Groups

There are numerous groups. Some groups look different, but actually have exactly the same structure and properties. Therefore, we need to sort them into classes according to their properties.

Basic classification properties of groups

(A) Order of a group - The **order** of a group **G** is the number of elements in **G**. Once the order of a group is found, we can quickly identify its possible structures.

Table: The order of a group and its possible structures

Order of a group	1	2	3	4	5	6	7	8	9	10	11	12	13	14
Possible structures	1	1	1	2	1	2	1	5	2	2	1	2	1	2

(1) From the table, we see that those groups of prime-number order have only one structure - cyclic.

(2) Groups from previous examples 2, 4, 5 and 6 look all "different", but they have the same order of 4. Their structures are either cyclic or Klein Four.

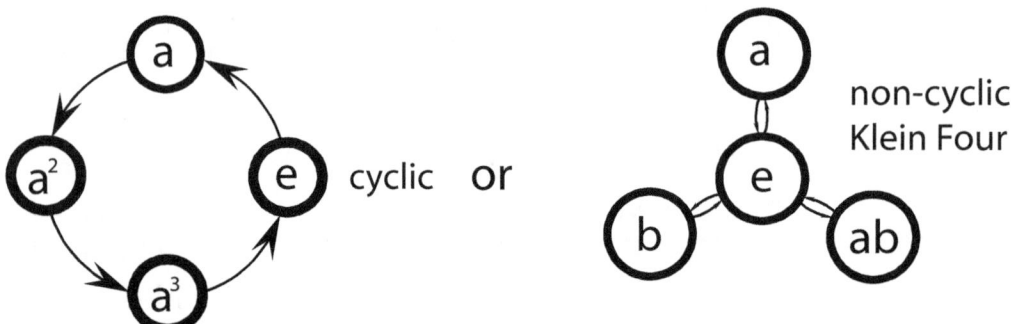

(3) There are two types of structures for groups of order 6.

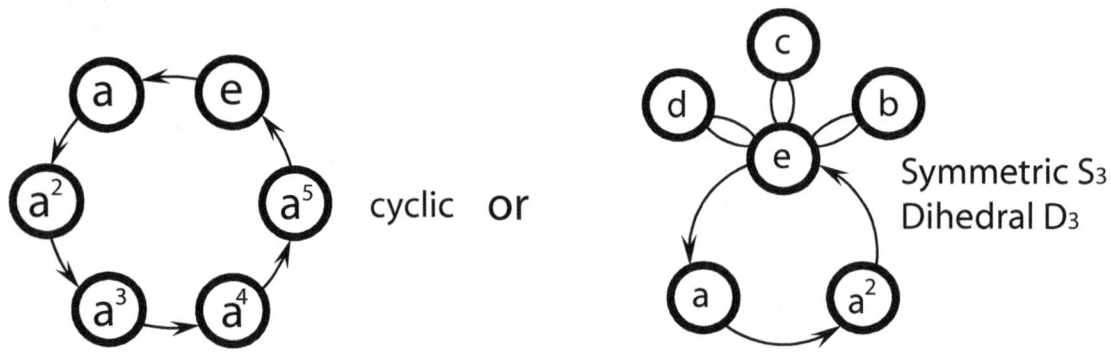

(Classification of Groups continued)

(B) Cyclic group - A group **G** is cyclic provided that **G** can be generated by some element in **G**. (See examples 2, 3 and 6)

(C) Commutative or Abelian group - A group is called *abelian* provided that its operation ∗ is commutative, $a * b = b * a$.

The commutative law states that when you add or multiply two numbers, you can reverse the order of operation without changing the result. For example, $2 + 3 = 3 + 2$, $4 \times 5 = 5 \times 4$ and $a + b = b + a$.

Non-Commutative Operations

Not all groups follow the law of commutativity, especially the groups in Symmetry and Quantum mechanics. Examples of non-commutativity can be found everywhere in everyday life.

Example 1

Wash your hands and then dry them

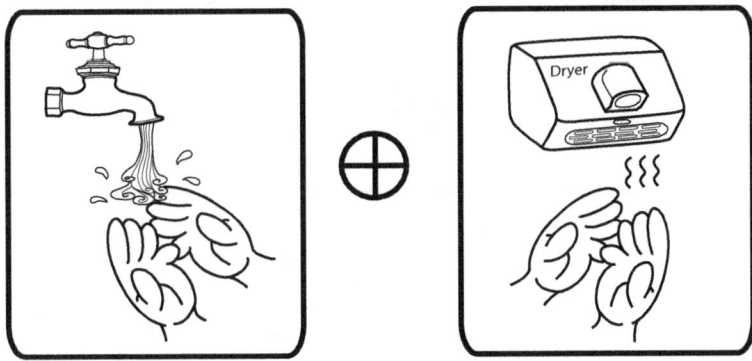

Dry your hands and then wash them

Example 2

The order of operation makes a difference.

Tooth extraction then anesthesia ≢ Anesthesia then tooth extraction

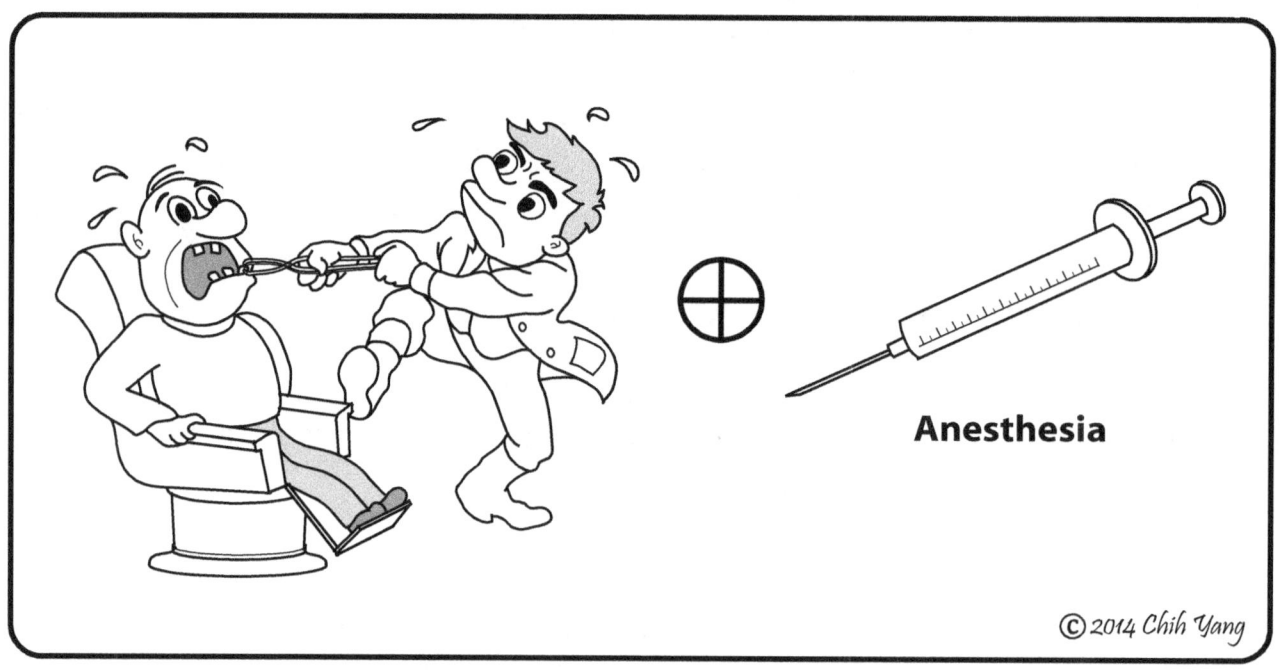

≢

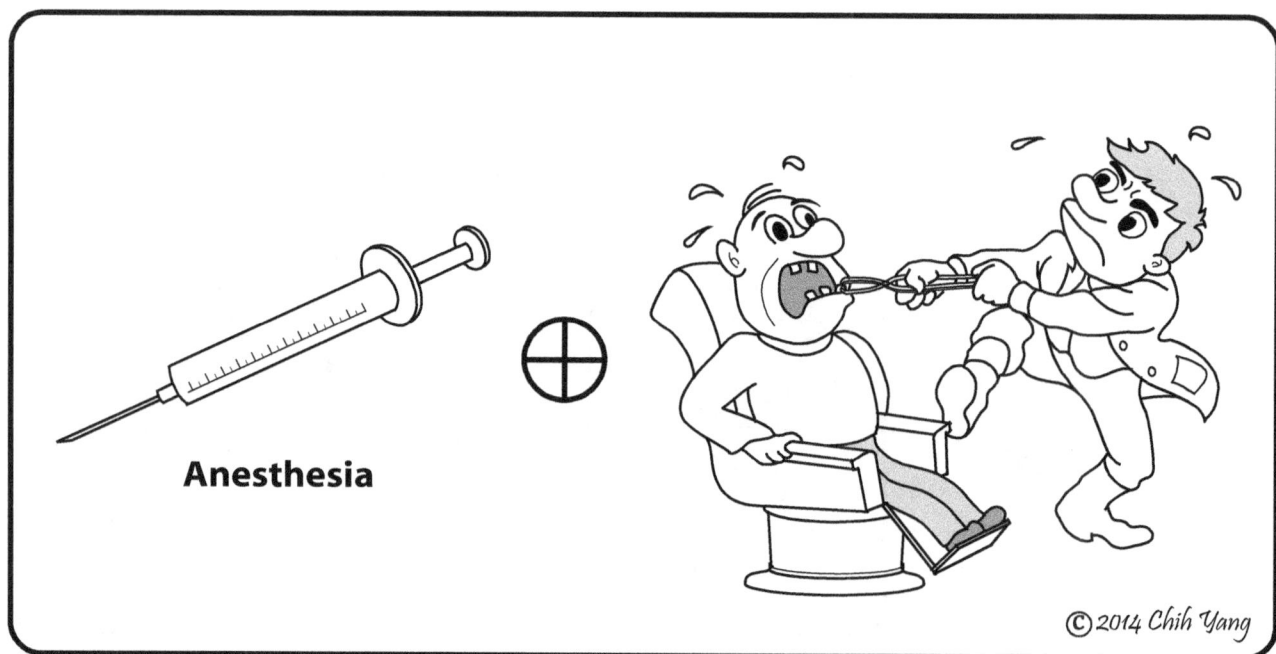

Example 3

Eat something sour, then something sweet ≨ Eat something sweet, then something sour

Sour ⊕ Sweet ≨ Sweet ⊕ Sour

Example 4

Dilute Sulfuric Acid

Sulfuric acid ⊕ Water ≠ Water ⊕ Sulfuric acid

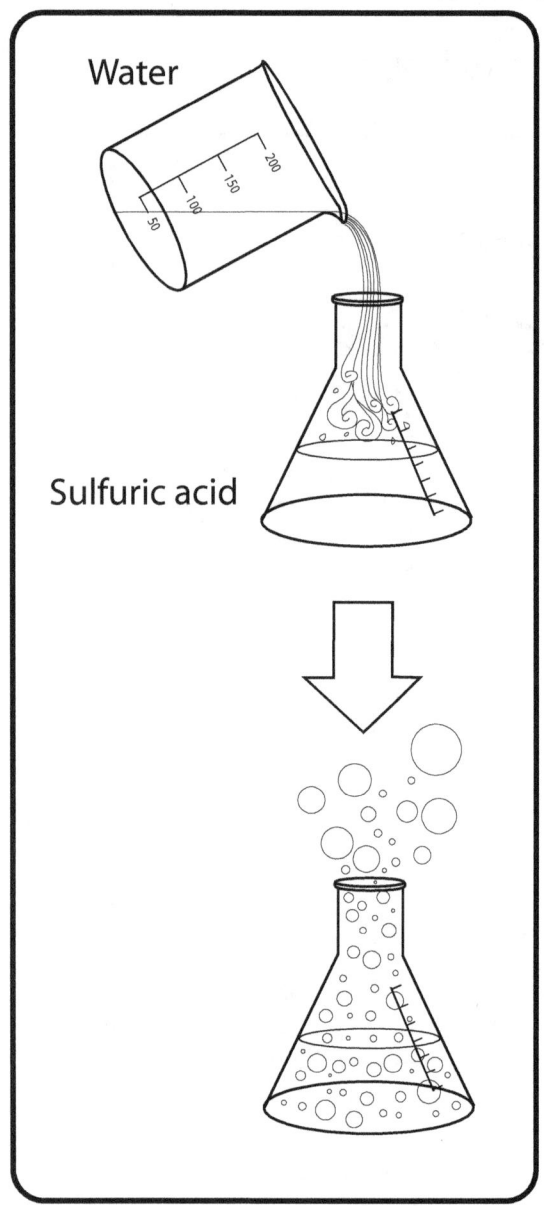
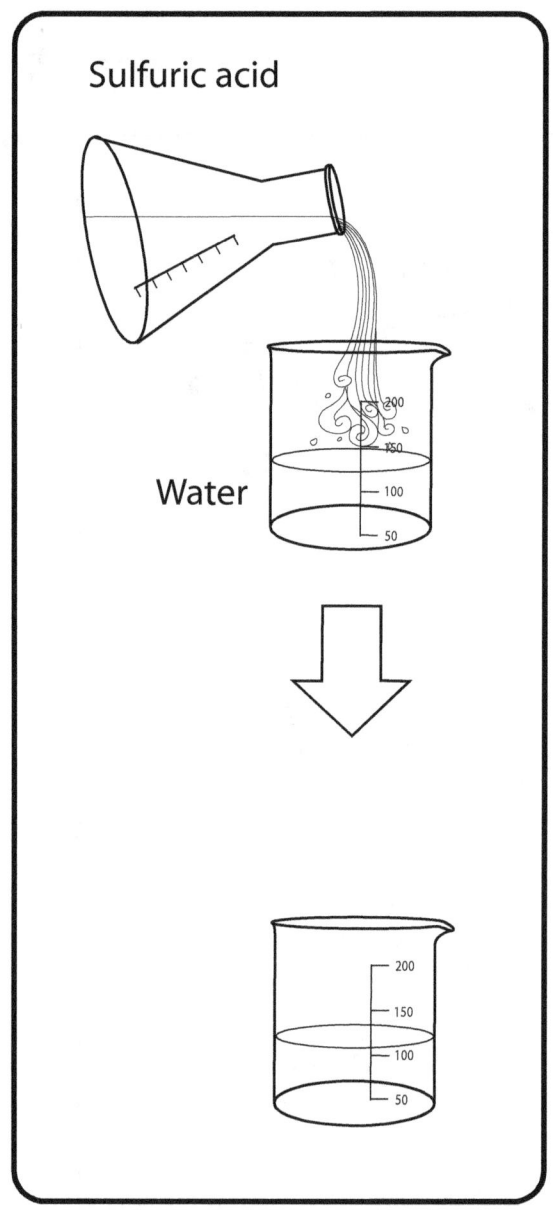

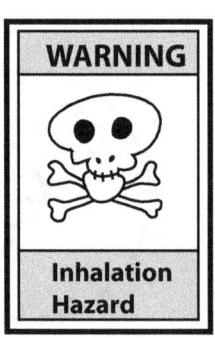

Example 5

Tires Rotations

Rotations are performed to optimize the life of tires.

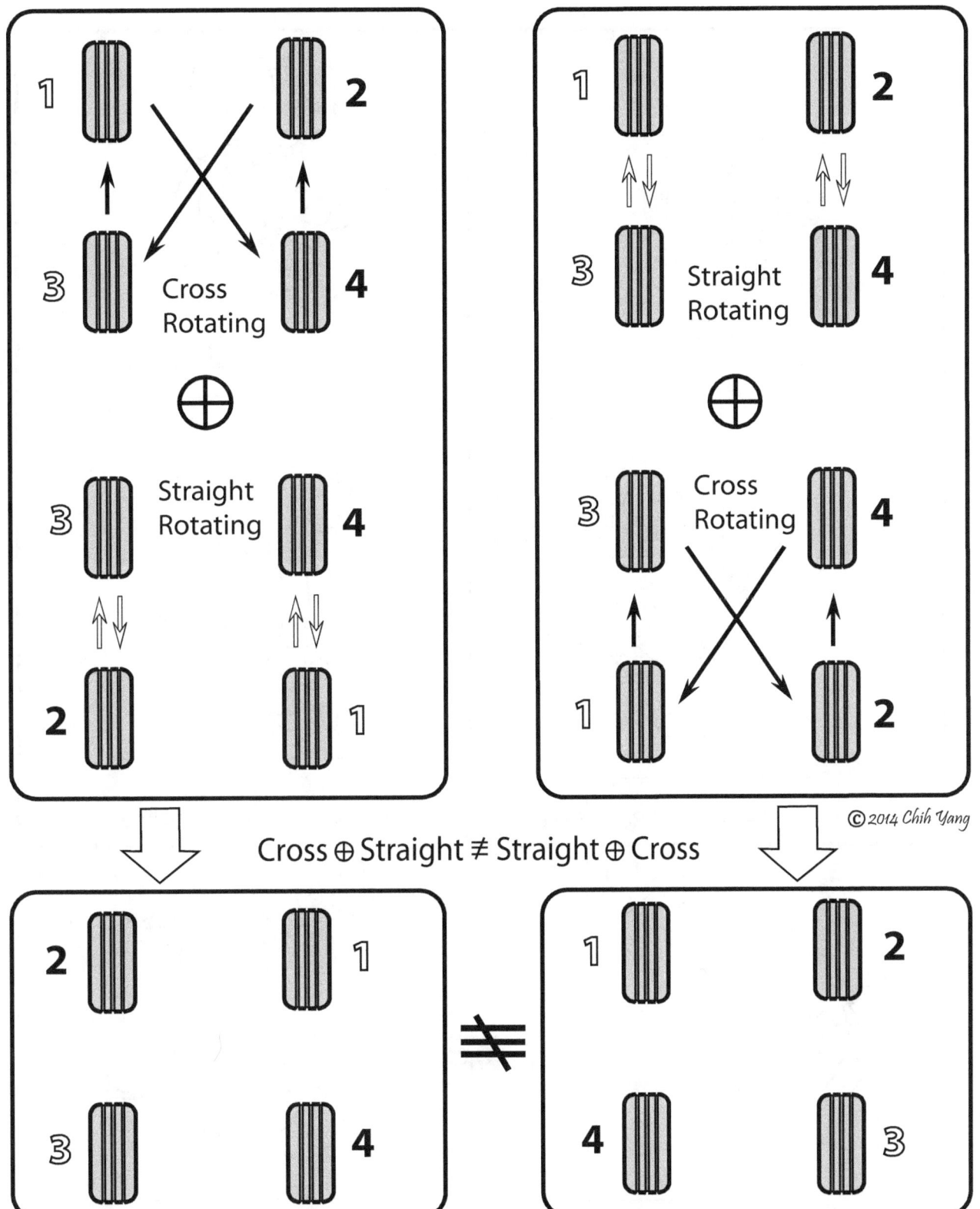

Cross ⊕ Straight ≠ Straight ⊕ Cross

© 2014 Chih Yang

Example 6

Dance
Incline ⊕ Rotate ≩ Rotate ⊕ Incline

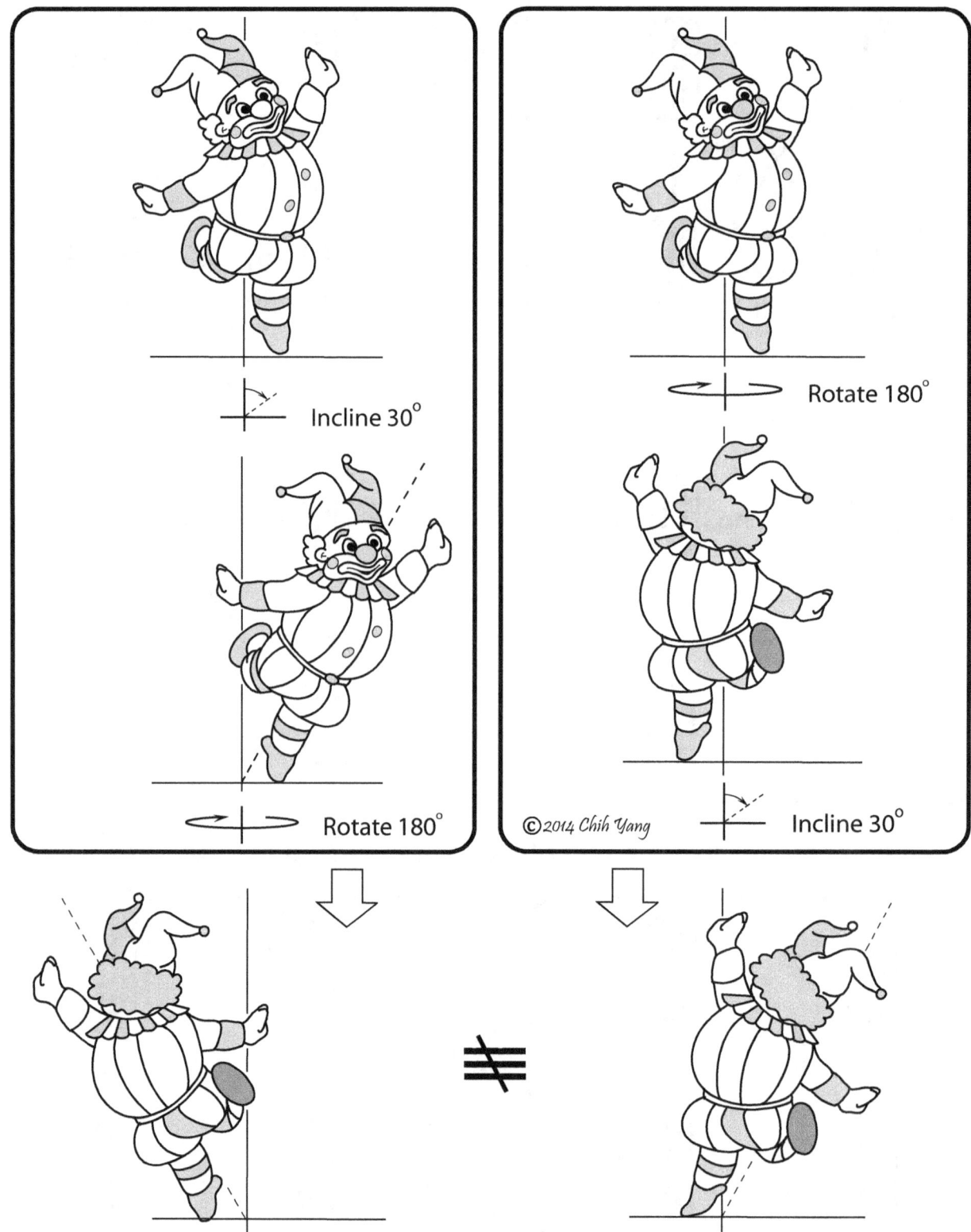

Applications of Non-commutativity

Example 1 The Quaternion Group

Let **G** be a set, **G** = { $1, i, j, k, -1, -i, -j, -k$}, where $1^2 = (-1)^2 = 1$, $i^2 = j^2 = k^2 = -1$, $ij = -ji = k$, $jk = -kj = i$, $ki = -ik = j$, and $-a = (-1)a = a(-1)$ for all a in **G**.

The rule of multiplication involving i, j, and k is described by the circle. This group has the following multiplication table.

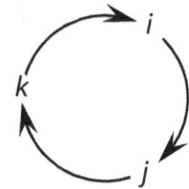

x	1	i	j	k
1	1	i	j	k
i	i	-1	k	-j
j	j	-k	-1	i
k	k	j	-i	-1

The operation is *non-commutative*.

G is a *non-Abelian* group with respect to multiplication. It was discovered by Sir William Rowan Hamilton in 1843 and applied to mechanics in three-dimensional rotations.

The Quaternion Group is useful in robotics, computer animation, computer vision, quantum physics, and crystallography.

Example 2 German Banknote

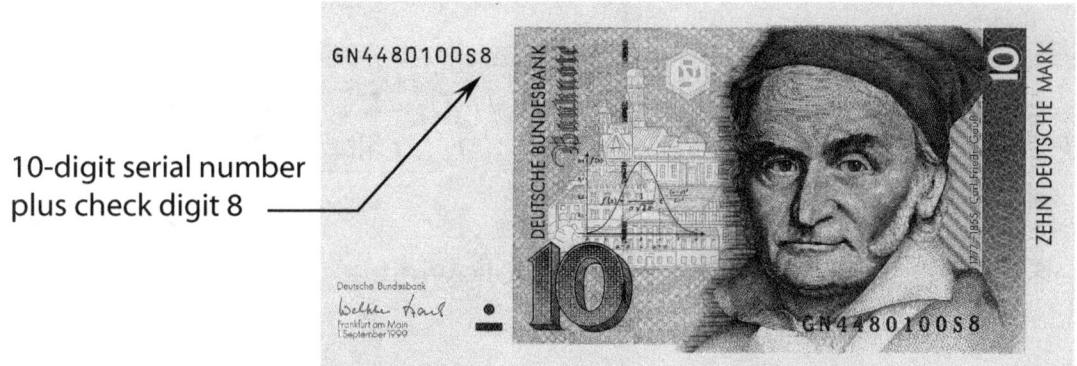

10-digit serial number plus check digit 8

Source: Wikimedia Commons (https://en.wikipedia.org/wiki/Deutsche_Mark#/media/File:DEU-10m-anv.jpg)

Check digits are used for error detection or security purposes in data coding. (see Chapter 7) There are several schemes available in computing the check digit. But they are not all infallible in detecting errors. For example, if a transposition error occurred, that error would not be detected.

To avoid this problem, the German government used Verhoeff's check-digit scheme to append a check digit to the serial number on German banknote (1990-2002). Verhoeff's scheme is a method based on the *dihedral group* D_5 that detects all transposition errors.

Dihedral Group D_5

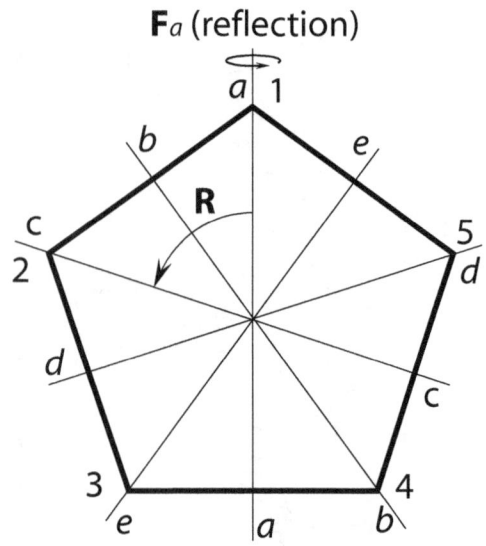

F_a (reflection)

Consider the pentagon to the left, let R denote the rotation of 72° counterclockwise. F_a denote the reflection about line *a-a*. The identity is denoted by *e*.

$$D_5 = \{e, R, R^2, R^3, R^4, F_a, F_aR, F_aR^2, F_aR^3, F_aR^4\}$$

Dihedral Group D_5 is a non-abelian group of order 10.

Isomorphism

Some groups may look outwardly different. If any two groups have the same algebraic structures, they are essentially the same.

Example 1 - Linear Transformation

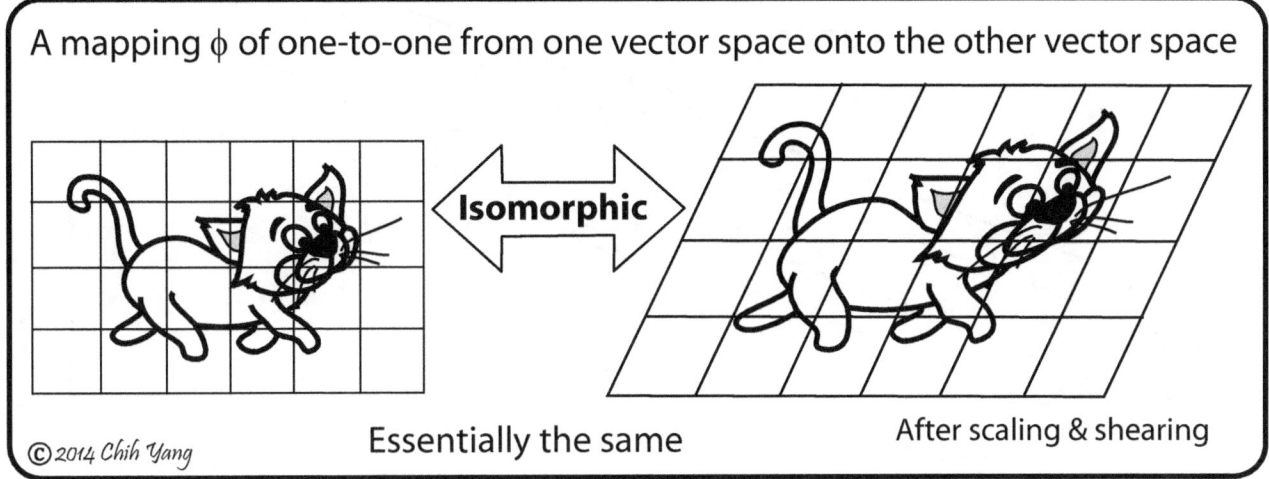

Two groups, **G** and **G'**, are essentially the same, if two conditions are met:

(1) There's an one-to-one mapping ϕ from group **G** onto **G'**.
(2) The mapping ϕ preserves the group operation, $\phi(ab) = \phi(a)\phi(b)$ for all $a, b,$ in **G**.

We say that the mapping ϕ is an **isomorphism**, and **G** is **isomorphic** to **G'**, $\mathbf{G} \cong \mathbf{G'}$.

Example 2 - Groups

Consider groups $\mathbf{G}=\{1,-1, i, -i\}$ under multiplication and $\mathbf{Z}_4 =\{[0], [1], [2], [3]\}$ under addition. Define a mapping $\phi: \mathbf{Z}_4 \rightarrow \mathbf{G}$ by the rule $\phi(x \cdot [1]) = i^x$, where x is any integer from 0 to 3.

The mapping ϕ is one-to-one and onto. It also preserves the operation:

$$\phi(x + y) = \phi((x + y)\cdot[1]) = i^{x+y} = i^x \cdot i^y = \phi(x) \cdot \phi(y)$$

Thus, group $\mathbf{Z}_4$ is *isomorphic* to **G**, $\mathbf{Z}_4 \cong \mathbf{G}$.

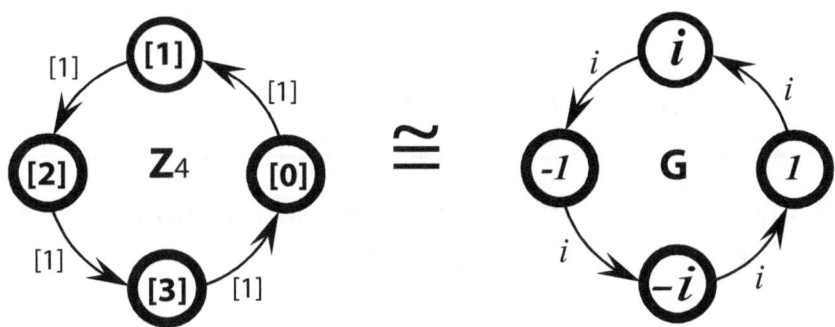

Isomorphism & Slide Rule

The slide rule is a mechanical analog calculator used primarily for multiplication, division and for functions such as roots and logarithms..

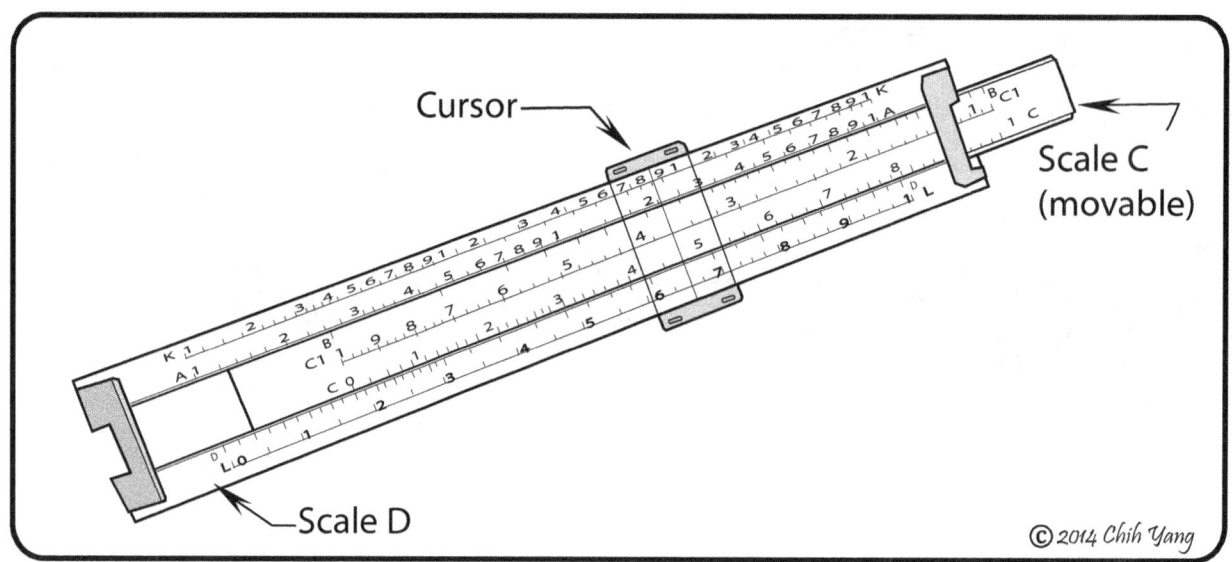

How it works:

To multiply 1.5 x 2 using the slide rule, place the 1 on Scale C directly above the multiplier 1.5 (see diagram). Then locate the second multiplier -2, on Scale C, and look directly below it to Scale D. There you will find the product -3.

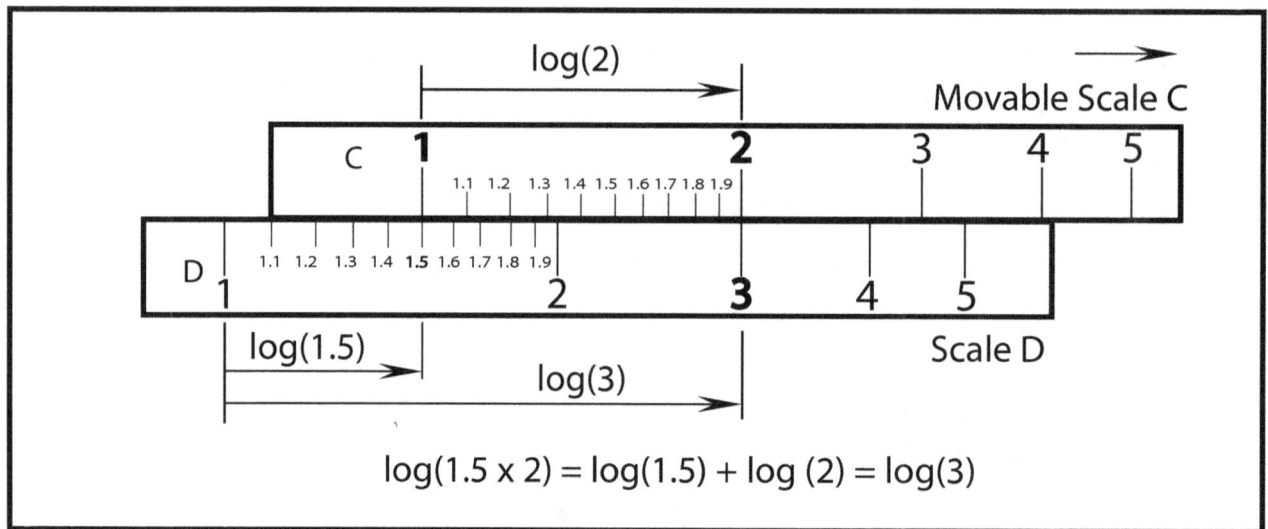

A slide rule works by transforming the operation of multiplication to the operation of addition. $\log(xy) = \log(x) + \log(y)$

By the use of logarithms, the operation of multiplication can be transformed to addition.

$$\log(xy) = \log(x) + \log(y).$$

The transformation is an isomorphism because $\mathbf{R^+} \cong \mathbf{R}$.

"The multiplicative group of positive real numbers $\mathbf{R^+}$ is isomorphic to the additive group of real numbers $\mathbf{R}$. The function $\psi: \mathbf{R^+} \to \mathbf{R}$ defined by $\psi(x) = \log(x)$, $\psi(xy) = \psi(x) + \psi(y)$, is an isomorphism".

When Noah's ark lands after The Flood, he lets all animals out and says, "Go forth and multiply."

Homomorphism

A homomorphism is a mapping ϕ from one group to another group, a mapping that preserves the operation but is not required to be one-to-one and onto. A homomorphism is a generalization of an isomorphism. A homomorphism links a group to its **image**.

Example 1

Consider the additive group of integer **Z** and a mapping $\psi: \mathbf{Z} \rightarrow \mathbf{Z}$ where $\psi(x) = 3x$ for all $x \in \mathbf{Z}$.

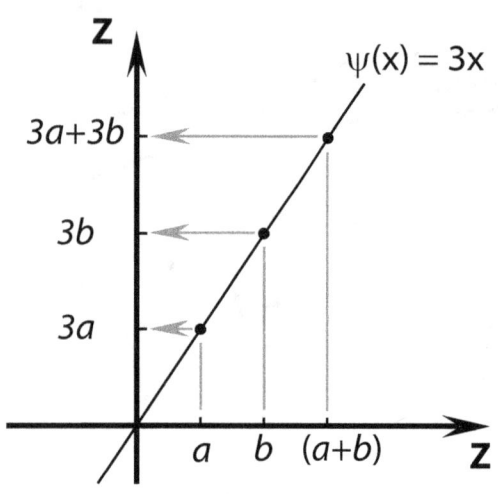

$\psi(a+b) = 3(a+b)$
$\quad\quad\quad = 3a + 3b$
$\quad\quad\quad = \psi(a) + \psi(b)$

ψ preserves the operation, is a homomorphism.
ψ is injective, one-to-one, but not surjective, onto.

Example 2

Consider the additive group of integer **Z** and the multiplicative group **G** = {$1, -1, i, -i$}, a mapping $\phi: \mathbf{Z} \rightarrow \mathbf{G}$ is defined by $\phi(x) = i^x$.

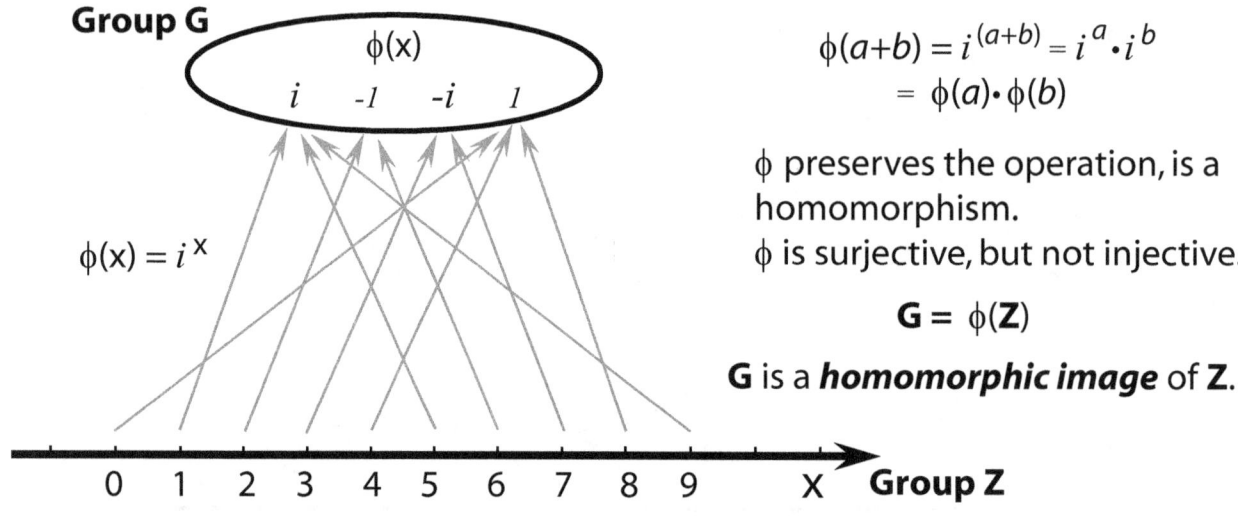

$\phi(a+b) = i^{(a+b)} = i^a \cdot i^b$
$\quad\quad\quad = \phi(a) \cdot \phi(b)$

ϕ preserves the operation, is a homomorphism.
ϕ is surjective, but not injective.

$$\mathbf{G} = \phi(\mathbf{Z})$$

G is a **homomorphic image** of **Z**.

In group theory, the main use of homomorphism is to create functions like $\phi: \mathbf{A} \to \mathbf{B}$ so that we can deduce things about **A** by looking at the properties (the algebraic structure) of its image **B**. It is like deducing a real-life object from looking at its photographic image.

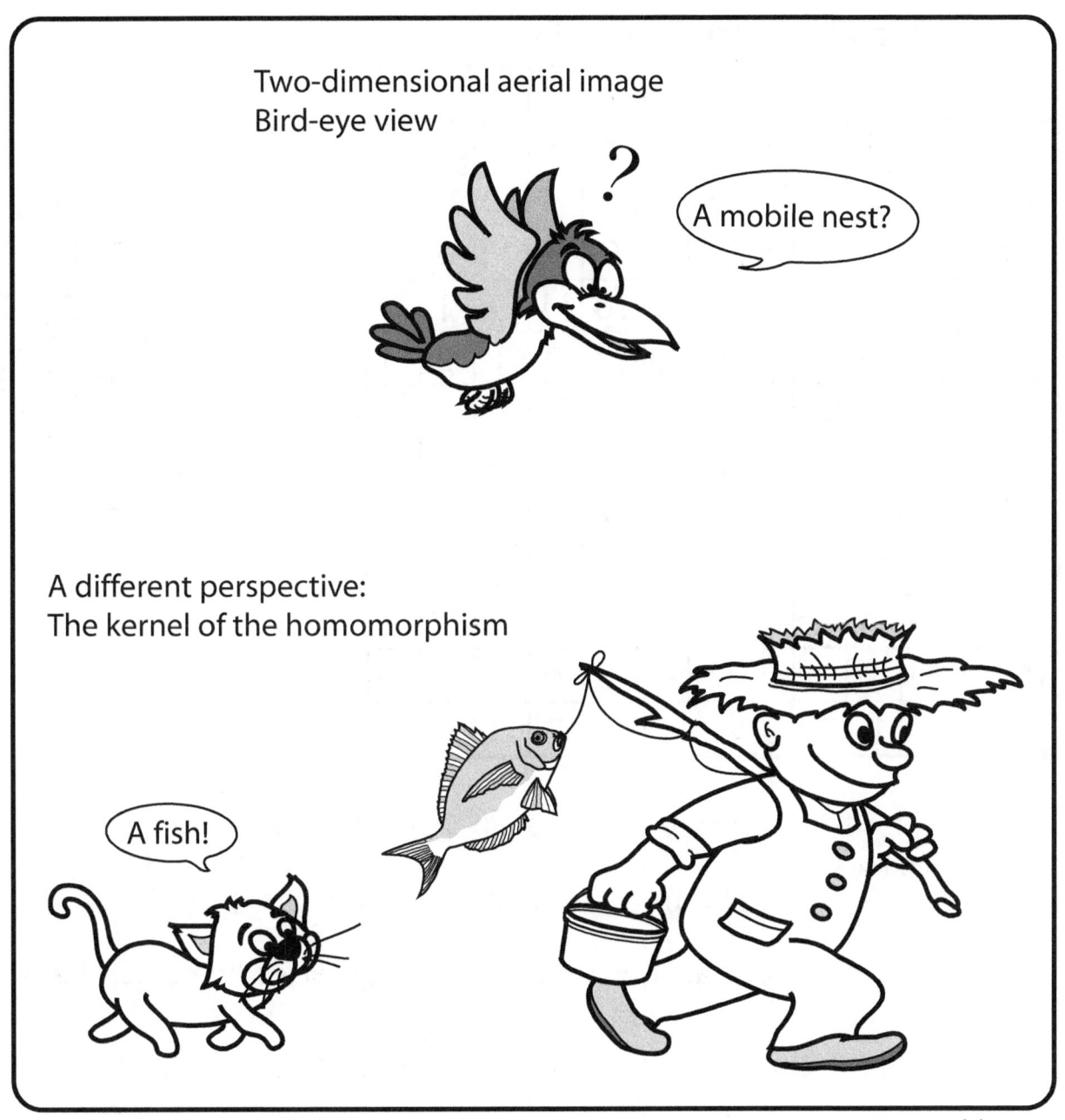

The kernel of the homomorphism reveals different features of an image by viewing it from a different perspective.

Example 3

For congruence class modulo 12, $\mathbf{Z}_{12}$, the function $\psi: \mathbf{Z}_{12} \to \mathbf{Z}_{12}$ defined by $\psi(x) = 3x$ is a homomorphism.

Cayley table for group $\mathbf{Z}_{12}$

+	0	1	2	3	4	5	6	7	8	9	10	11
0	0	1	2	3	4	5	6	7	8	9	10	11
1	1	2	3	4	5	6	7	8	9	10	11	0
2	2	3	4	5	6	7	8	9	10	11	0	1
3	3	4	5	6	7	8	9	10	11	0	1	2
4	4	5	6	7	8	9	10	11	0	1	2	3
5	5	6	7	8	9	10	11	0	1	2	3	4
6	6	7	8	9	10	11	0	1	2	3	4	5
7	7	8	9	10	11	0	1	2	3	4	5	6
8	8	9	10	11	0	1	2	3	4	5	6	7
9	9	10	11	0	1	2	3	4	5	6	7	8
10	10	11	0	1	2	3	4	5	6	7	8	9
11	11	0	1	2	3	4	5	6	7	8	9	10

$\psi \Rightarrow$

Image $\psi(\mathbf{Z}_{12})$

+	0	3	6	9
0	0	3	6	9
3	3	6	9	0
6	6	9	0	3
9	9	0	3	6

$$\psi(x) = 3x = \begin{pmatrix} x= & 0 & 1 & 2 & 3 & 4 & 5 & 6 & 7 & 8 & 9 & 10 & 11 \\ 3x= & 0 & 3 & 6 & 9 & 0 & 3 & 6 & 9 & 0 & 3 & 6 & 9 \end{pmatrix}$$

ψ preserves the operation, $\psi(x + y) = \psi(x) + \psi(y)$. ψ is a homomorphism.

Cosets: {0, 4, 8}   {1, 5, 9}   {2, 6, 10}   {3, 7, 11}
              ↓↓↓   ↓↓↓   ↓↓↓   ↓↓↓
Images: 0   3   6   9

Kernel φ is the set of all elements carried to the identity by ψ.

Since 0, 4 and 8 are the elements carried by ψ to the identity 0, Kernel φ = {0, 4, 8}.

Following cosets and their corresponding images, the Cayley table of group **Z**₁₂ is rearranged as below.

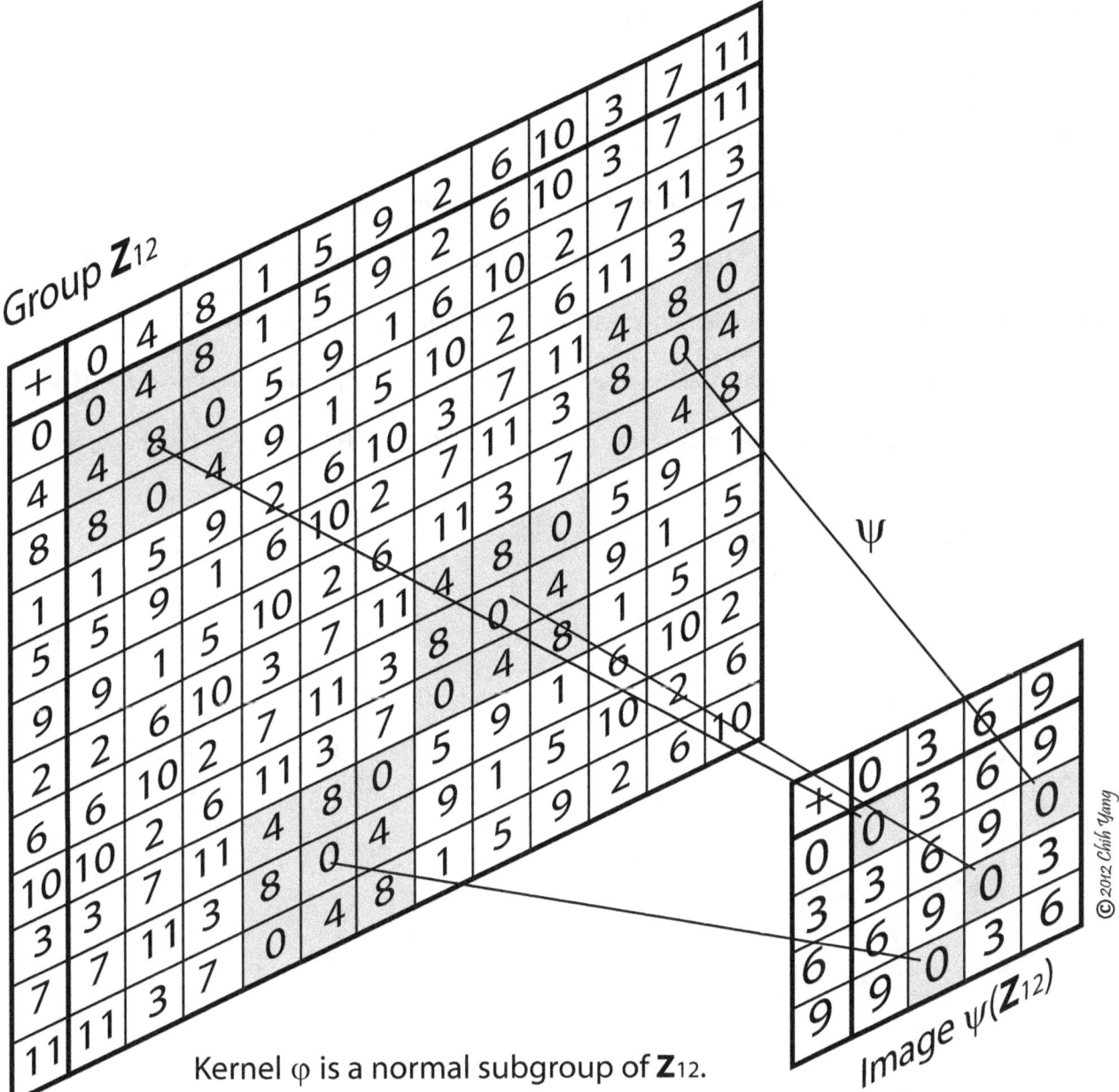

Kernel φ is a normal subgroup of **Z**₁₂.

Kernel φ = {0, 4, 8} views ψ from a different angle, reveals more features and leads to further topics of factor groups.

References – Algebraic Systems

[1] Ash, Robert B.; A Primer of Abstract Mathematics, The Mathematical Association of America, Washington, DC, 1998

[2] Bajnok, Bela; An Invitation to Abstract Mathematics, Springer, 2013

[3] Bloch, Norman J.; Abstract Algebra with Applications, Prentice-Hall, Inc., NJ, 1987

[4] Berlinghoff, William P.; Mathematics-the Art of Reason, D.C. Heath and Company, Boston, 1968

[5] Gallian, Joseph A.; Contemporary Abstract Algebra, 8th edition, Cengage Learning, 2013

[6] Grossman, Israel and Wilhelm Magnus; Groups and Their Graphs, Random House, 1964

[7] Lidl, Rudolf and Gunter Pilz; Applied Abstract Algebra, 2nd Edition, Springer, 1997

[8] Maxwell, E. A.; A Gateway to Abstract Mathematics, Cambridge at the University Press, UK, 1965

[9] Pinter, Charles C.; A Book of Abstract Algebra, 2nd Edition, McGraw-Hill Publishing Company, NY, 1990

[10] Weyl, Hermann; Symmetry, Princeton University Press, 1952

Appendix

A Pattern from Finite Differences

- an example of data smoothing in finite differences.

> One of the following numbers was misprinted, which one?
>
> **1   3   6   11   20   31   48   71   101**
>
> Hint: There is a pattern when you reach the fourth differences.

The differences table is the standard format for displaying finite differences.

		1		**3**		**6**		**11**		**20**		**31**		**48**		**71**		**101**
First differences →			2		3		5		9		11		17		23		30	
				1		2		4		2		6		6		7		
						1		2		-2		4		0		1		
Fourth differences →							1		-4		6		-4		1			

The error increases for higher differences and has a binomial coefficient pattern.

Changing **20** to **19** brings the new table:

1	**3**	**6**	**11**	**19**	**31**	**48**	**71**	**101**
	2	3	5	8	12	17	23	30
		1	2	3	4	5	6	7

The pattern suggests that number 20 was misprinted. The correct number should be 19.

Appendix

A Well-defined Binary Operation

Example 2

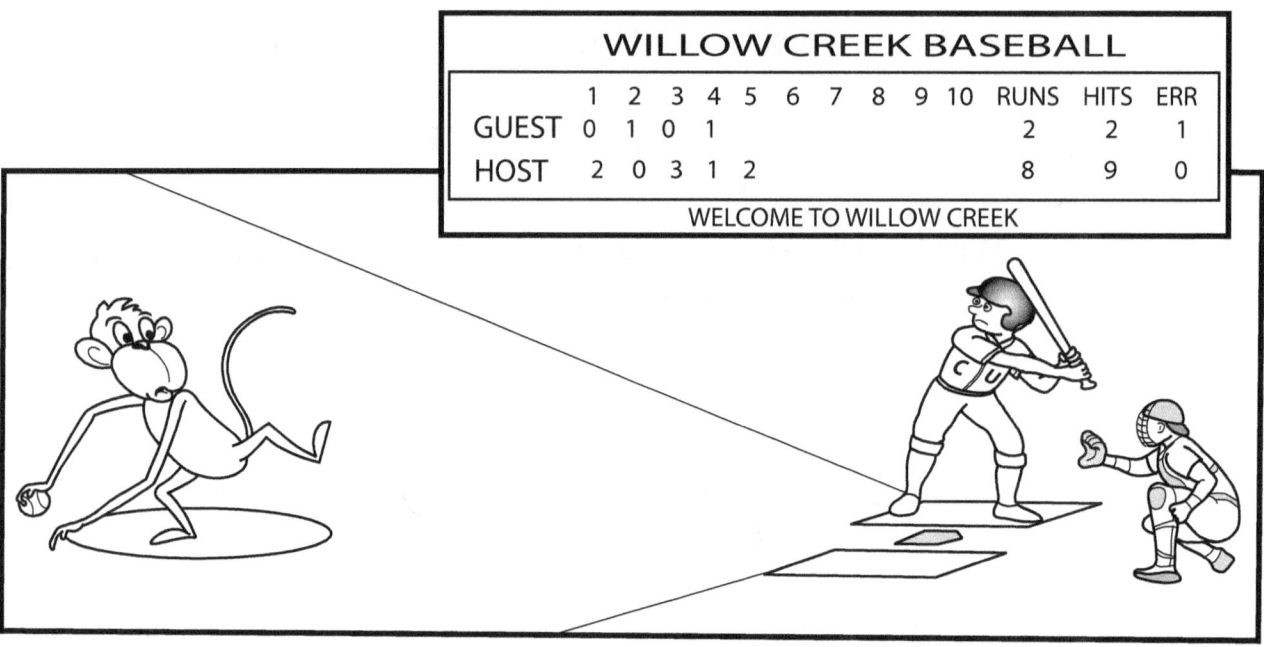

The Willow Creek baseball team has a game record for the past three seasons of 7 wins out of 13, 6 wins out of 12 and 8 wins out of 14 games.

Add them up:
$$\frac{7}{13} \oplus \frac{6}{12} \oplus \frac{8}{14} = \frac{21}{39}$$

There is a total of 21 wins out of 39 games.

The numbers 7/13, 6/12 and 8/14 are not rational numbers.

$$\frac{6}{12} \neq \frac{1}{2} \quad \text{and} \quad \frac{8}{14} \neq \frac{4}{7}$$

It works because the operation $\oplus$ has only one image.

Appendix
Bomber's Bullet Holes Pattern

- During World War II the British bombed the Germans daily.
- Wanting to improve the odds of bringing troops home, a mathematician named Abraham Wald studied the location of bullet holes in British planes and determined where to added armor plating.

Where would you add the armor plating to?

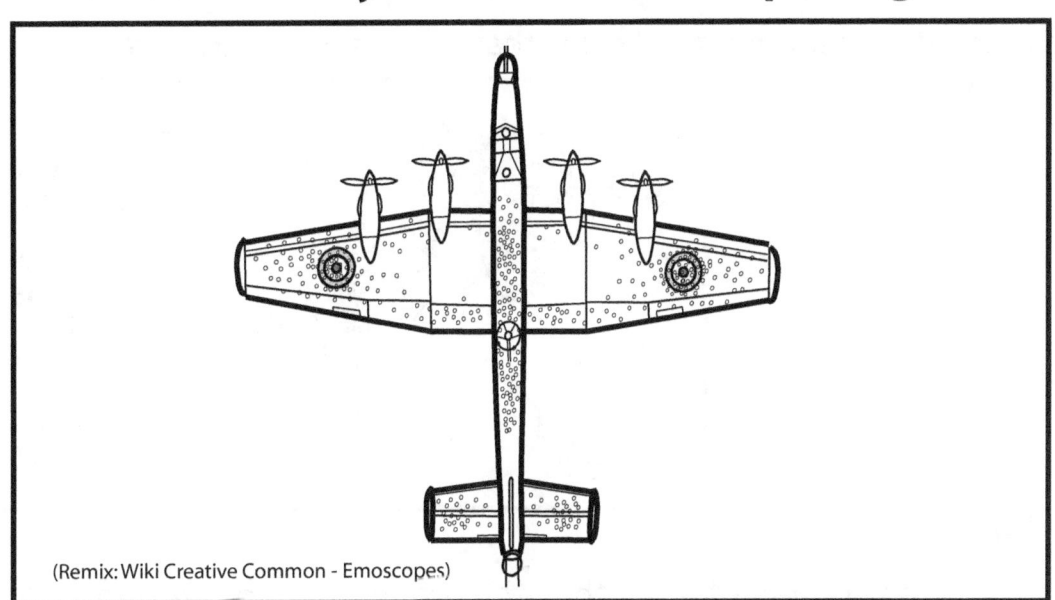

(Remix: Wiki Creative Common - Emoscopes)

Add the armor to the spots with no bullet holes.
These planes were hit in area and were able to return, whereas those that did not make it back had been hit in more critical points.

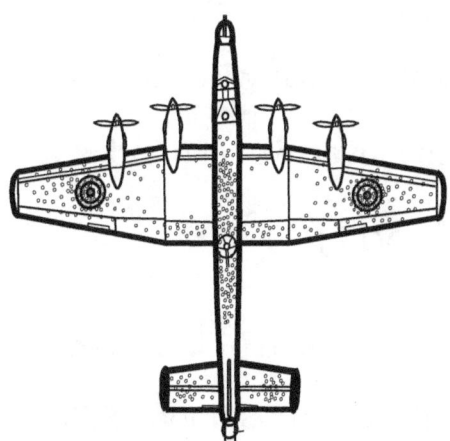

The plane that made it back

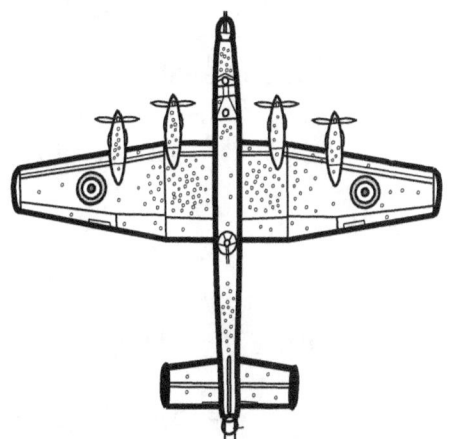

The plane that did not return

Appendix

Pirates of the Carolinas

A band of 17 pirates wanted to divided their loot of gold coins evenly. When they divided the coins into equal piles, 5 coins were left over. In a brawl over who should get the extra coins, two of the pirates were slain. Again, the remaining 15 pirates tried to divide the loot equally, 3 coins remained. Another pirate was killed in the fighting again over the extra coins. Finally the last 14 pirates divided the loot equally among themselves.

What was the least number of coins that the pirates could have been distributing?

Solution:

Let x be the number of coins, then the problem can be written as:

$$\begin{cases} x = 5 \pmod{17} \\ x = 3 \pmod{15} \\ x = 0 \pmod{14} \end{cases}$$

Where $R_1 = 5$, $R_2 = 3$, $R_3 = 0$, $m_1 = 17$, $m_2 = 15$, and $m_3 = 14$,

For the system of equations, m_1, m_2 and m_3 are pairwise coprime.

Let $M = m_1 \cdot m_2 \cdot m_3$, $b_i = M/m_i$ and b_i^{-1} be the inverse of $b_i \pmod{m_i}$.

$$x = \sum_{i=1}^{3} R_i \, b_i \, b_i^{-1} \pmod{M}$$

Since $b_1 = 210$, $b_2 = 238$, $b_3 = 255$, $b_1^{-1} = 3$, $b_2^{-1} = 3$, and $b_3^{-1} = 5$, then

$$x = (5)(210)(3) + (3)(238)(7) + (0) = 8148 = 1008 \pmod{3570}$$

The least number of coins is 1,008.

Appendix
RSA Public-key Cryptosystem

Cryptographic Key Generation

RSA cryptosystem is widely used for secure data transmission. A user needs two keys, one public key and one private key.

Example of Making the keys

Two prime numbers p and q are chosen as the private key:

Let $p = 23$, and $q = 43$

Then we make the public key from p and q.

$m = p\,q = 23 \times 43 = 989$
$(p-1)(q-1) = 22 \times 42 = 924$
Find a number **e**, which must be relatively prime to $(p-1)(q-1)$

We choose $e = 221$, which to relatively prime to 924.

Public key

$$m = 989,\ e = 221,$$
$$\text{encrypt equation}\quad f(x) = x^{221}\ (mod\ 989)$$

to be made public

Since $ed = 1\ (mod\ (p-1)(q-1))$

solve the equation $221d = 1(mod\ 924)$, we found $d = 485$.

Private key

$$p = 23,\ q = 43,\text{ and }\ d = 485$$
$$\text{decrypt equation}\ g(x) = x^{485}\ (mod\ 989)$$

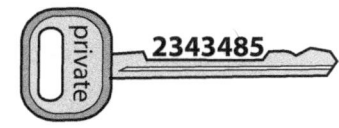

keep secret

Index

A
Abel, Niels Henrik, 7, 88
    Abelian group, 189
abstract thinking, 25
    abstraction, 52
Achilles, 52
achiral, 167
aesthetics, 159
aleph zero, 57
algebraic laws, 178
algebraic systems, 178
algorithm, Euclidean, 115 - 118
al-Khwarizmi, Muhammad ibn Musa, 86
Aristitle, 154, 155
    Aristorelian view, 41
    Aristotelian reasoning, 154
associative law, 178
asymmetry, 167
axiomatic formulation, 66

B
Bach, Johann Sebastian, 173
Bacon, Francis, 156
bank identification numbers, 126
Bernoulli, Johann, 9
big dipper, 171
bioinformatics, 20
Boolean function, 31

C
Cantor, Georg, 51
casting out 9s, 128
calculus, 8
cardinality, 56
Cardano, Girolamo, 6, 87
Cayley table, 182
chirality, 167, 168, 169
clock arithmetic, 110
closure law, 178
    closed, 95
classification of groups, 188
cloud computing, 15
codomain, 91
commutative
    law, 179
    group, 189

concrete thinking, 25
congruent, 110, 120, 121
    congruence of integers, 120
continuum, 64
conservation of momentum, 170
cosets, 202
counting, 134
    Chinese remainder, 135
cryptography, 144
    public-key, 145
    private-key, 144
    RSA cryptosystems, 146
cut, Dedekind's, 63
cyclic group, 183
cycloid, 9

D
Dedekind, Richard, 62
    Dedekind's cut, 63
Descarte, Rene, 156
dihedral group, 196
Dirichlet, Peter Gustav Lejeune, 90
distributive law, 179
DNA sequence, 20
dodecahedron, 181
domain, 91

E
Einstein, Albert, 43, 159, 160
Euclid, 3, 24
    Euclidean algorithm, 115 - 118

F
field, 180
finite difference, 21
Ford assembly line, 35
fractal, 10
function, 89 - 94

G
Galilei, Galileo, 41, 155
Galois, Evariste, 89, 89
Gauss, Carl Friedrich, 70, 121
German banknote, 196
German tank problem, 14
Gershon, Levi ben, 75

golden ratio, 118
Golomb's tromino theory, 79, 81
Gordian knot, 10
greatest common divisor, 114
group, 180
- abelian, 189
- classification, 188
- commutative, 189
- crystallographic groups, 166
- cyclic, 183
- dihedral, 196
- drill commands, 187
- factor, 203
- frieze groups, 162
- generator, 182
- Klein four, 184
- non-frieze groups, 163
- order, 183
- permutation, 181
- space groups, 166
- wallpaper groups, 1

groundhog day, 91

H
Hamilton, William Rowan, 195
Hanoi, tower of, 76
Hilbert's hotel, 58
Hindu-Arabic numerals, 4, 86
homochirality, 168
homomorphism, 200
- homomorphic image, 200
- kernel, 201

I
Icosahedron, 181
identity, 178, 179
inertia, the law of, 41, 42
inverse, 178
infinity, 52
- infinity monkey, 46
integral domain, 180
image, 91
imaginary number, 87
image recognition, 174
invariant, 159, 161, 162
irrational number, 119
isomorphism, 197
- isomorphic, 186

K
Klein four group, 184
knot theory, 10
Koch's snowflake, 10

L
Leibniz, Gottfried Wilhelm, 8, 61
Leonardo da Vinci, 159
Lucas, E., 76

M
mapping, 90
mathematical induction, 71
- the principle, 73
Michelson-Morley experiment, 157
mining the data, 19
modular arithmetic, 110, 114
- modulus, 120, 121

N
natural numbers, 54
negative numbers, 5
Newton, Isaac, 8, 156
Noether's theorem, 170
non-commutativity, 189
- non-commutative operations, 189

O
operations
- binary, 89
- unary, 100
order of group, 183
octahedron, 181

P
Pachelbel canon in D, 173
Panther tank, 14
paradox
- barber's, 65
- liar, 66
- Russell's, 65
- Zeno, 52
Pasteur, Louis, 167
patent law, 29
permutation group, 181
Plato's allegory, 37
Platonic solids, 181
precision arithmetic, 142

Q
quantum mechanics, 160
quaternion group, 195

R
rational numbers, 60
real numbers, 61
relation, 89, 101
    directed graphs, 103
    equivalence, 108
    reflexive, 105
    symmetric, 105
    transitive, 108
relativity
    Galilean, 42
    Einstein, 43, 160
ring, 180
Rubik's cube, 181
Russell, Bertrand, 65

S
SENTRI, 17
set theory, 50
    Zermelo-Fraenkel, 66
Shakespeare, William, 26, 46
Sherman tank, 14
slide rule, 198
square pyramid numbers, 74
statistical estimator, 15
subset, 56
symmetry, 159, 161

T
tetrahedron, 181
    tetrahedral point group, 181
time dilation, 43
Thalidomide, 169
thought experiment, 37
transformation, 90
transposition errors, 129
tromino puzzle, 79
triangle numbers, 73
two-way switch circuit, 185

U
unary operation, 100
UPC symbols, 125

V
Verhoeff's check-digit scheme, 196
virtual reality, 47
Vitruvian man, 159

W
Wald, Abraham, 16
well-defined, 95
Weyl, Herman, 159
word size, 142
Wright brothers, 29

Z
Zeno's paradox, 52
Zermelo-Fraenkel set theory, 66
zero, 5
zero divisors, 179

www.ingramcontent.com/pod-product-compliance
Lightning Source LLC
Chambersburg PA
CBHW080410300426
44113CB00015B/2461